Configuration Management for Senior Managers

Configuration Management for Senior Managers

Essential Product Configuration and Lifecycle Management for Manufacturing

Frank B. Watts

AMSTERDAM • BOSTON • HEIDELBERG • LONDON
NEW YORK • OXFORD • PARIS • SAN DIEGO
SAN FRANCISCO • SINGAPORE • SYDNEY • TOKYO

Butterworth-Heinemann is an imprint of Elsevier

List of Figures

Preface

Engineering documentation control (EDC), configuration management (CM), and product lifecycle management (PLM) are names and acronyms given to what this author considers to be much the same subject—thus the shortened title—*CM for Senior Managers*.

CM for Senior Managers is about the important issues and policies in the product life cycle for *product* manufacturing companies. CM is, without doubt, a critical discipline for any size operation because the company's product is embodied in the design drawings and specifications.

We will identify what is important for the executives, senior managers, and others to understand about the control of the product design; electronics, mechanics, hydraulics, fluids, and embedded software documentation throughout the product's life.

Most of the critical issues are common for companies whether they are a make to stock, make to print, make to order, design to order, or combinations of these. Most issues are the same whether the product is simple or complex. This text will generally be directed at the make to stock company because it is probably the broadest CM environment. Some "translation" will be required of the reader in the other modes of operation.

The software CM (small s) discussed in this book is *not* to be confused with software configuration management (SCM) wherein the organization's product *is* "software."

SCM is what you will find most frequently if you Web search "CM." That is also what your software engineering folks may want to have to control the design and development of your product's embedded software or internal process software tool.

This book will include discussion of, and is applicable to, software when it is part of the product—just as mechanics, electronics, etc. are embedded in the product. This book will not be about SCM, just as it won't be about CAD or other design tools.

Most executives in product manufacturing realize that their new product design and release process is slow, confusing, and often error-prone. They understand that the change process is bogged down in a multitude of requests, changes, departments, functions, systems, signatures, personalities, and that it is slow, confusing, and perhaps divisive.

Execs may also realize that they have more than one bill of material for every product but usually aren't completely sure why. They know that there are many departments involved in those product life cycle processes and that there is considerable finger-pointing going on, but senior management often isn't quite sure what to do about it. This book will help if not cure.

Critical metrics/key performance indicators will be included in each applicable chapter.

Looking up from the indirect workers' viewpoint there is a different perspective. During my 33 years of experience in/with product manufacturing companies (including stints at the director level) and another 22 years of consulting with such companies, one management characteristic stands out among the engineers, technicians, managers, and others close to the CM System: "The senior management doesn't understand or appreciate what we do!" I've heard that cry from CM managers, their managers, and their people in almost every company I've worked for and with (over 75 total). Sometimes even the chief engineer sounds the same tune about his/her management.

Some of this attitude is based upon a lack of knowledge as to what the senior management actually does know about the discipline. Based upon my personal experience the senior management knows enough about it to know it is fraught with problems and to wish it would just go away—be transparent. I often hear management say: "The people over there just aren't cutting it!" When I talk to those people I hear: "The management doesn't understand." This writer's experience says that it usually isn't the people or the management—it is usually the **processes that are the problem**.

The paradox here is that it can be made to "go away"—if good basics are put in place—the discipline will be largely transparent to executive management. **With best of the best practices in place technical folks in engineering and operations will both set free to innovate.**

Until that occurs, the top-down and bottom-up views will be quite different, and the finger-pointing will continue.

There are plenty of challenges facing senior management in product manufacturing. Often when a problem/challenge becomes apparent, there is a tendency to leap quickly to search for a new software application as the solution. This writer calls this current phenomenon "app mania." This trend is painfully obvious to an outsider and painfully expensive and frustrating for many an insider.

Much can and should be done with legacy software and manual processes before purchasing the next software "solution"! What should be done instead … is exactly what this book is about.

The term "configuration management" was invented by the DOD folks. The DOD version of CM is very convoluted and complex but need not be. The best of the best management policy, practices, and process guides found here are applicable to both military and commercial products.

Whether you think of the discipline as configuration management (CM), product life cycle management (PLM), or engineering documentation control (EDC), the same basic management policies should apply. However, recognize that the DOD, IT, and other folks have brought more than a little distortion to those terms. Nevertheless, from a basic configuration management perspective CM, PLM, EDC are essentially interchangeable terms. The distortion injected by the software and DOD folks can and will be largely ignored—except when unavoidable.

One writer describes the discipline as being "highly technical in nature." This is not surprising since that writer was from an information technology/military background—wherein both arenas have made the discipline complicated, confusing, and complex—but that doesn't make it "highly technical."

The bill of material (BOM) portion of ERP and PLM systems will also be considered part of the same discipline/problem/challenge in this text.

Many references to actual company experiences will be included but generally without reference to the company name—since the author has signed nondisclosure agreements with most clients.

The primary purpose of this book is to give executive and senior management an appreciation for the importance of the discipline, examples of good and bad practices, an understanding of the essential elements, as well as an outline of the role which they should play in it. It need not be complicated. The author likes to think of himself as the "Vince Lombardi of CM" and takes pride in reducing the discipline to blocking and tackling fundamentals.

This book will hopefully bring this basic blocking and tackling approach to *CM for Senior Managers*—no software "app mania," no use of "BS Bingo," IT "clouds," "augmented reality," "paradigm shifts," "single points of truth," "granularity," or other consultant's obfuscation will be employed. Minimal repetition will be used in an attempt to make the chapters on the basic processes to be stand-alone or to emphasize critical policy. Minimal use of basic acronyms and only basic policy for maximizing company profits will be included.

All of the policies, rules, and ideas in this book do not need to be implemented for best in class results. A few can be ignored with very good results. But if best in class CM processes are to exist, most of them need to be heeded.

As Peter Drucker wisely stated: "Efficiency is doing things right. Effectiveness is doing the right things." Executive policy, rules, and practices will be developed in this book to effectively and efficiently guide the CM/PLM/EDC processes and bill of material (BOM) discipline in your company.

Policies and Critical Practices

POLICIES

If best-in-class processes are to be attained, every company/division should have one executive committed to be the CM champion. **2**

The product design must be documented and controlled effectively and efficiently for profitable, sustainable, production, service, and sale of our products. **5**

CM must be chartered, manned, and expected to be both communicator and the conveyor of new design documents and changes to the right people at the right time. **7**

One CM office shall serve all the projects in one logical business unit. **10**

One CM "organization," usually in Engineering, needs to be established—part of a person in a start-up, a person as you grow, and several people when successful. **11**

When your operations are multiplant, a slim corporate function with each plant/division having a CM function is appropriate. **11**

The CM organization should answer directly to the Chief Engineer or to the Director of Engineering Services—not any lower in the organizational chart. **12**

ISO certification must be recognized as only the first step out of chaos. Standards implementing the best of the best practices should follow. **13**

Satisfy good commercial CM practices seeking the best of the best processes, then look at modifications to satisfy unique Mil Spec requirements. **14**

The first "best-in-class" general standard written should outline the responsibilities of the CM organization. **15**

CM must have control of rev levels after release, or you do not have control. **18**

Any process designed by man can be improved by man. **19**

CM processes with current software shall be addressed first, measured, and brought to a reasonably fast, well understood, efficient, effective, minimally controlled state, and then automated for efficiency if necessary. **24**

Put any proposals for purchase of new—or major changes to—ERP, CM, PDM, PLM software on hold until the CM processes all measure up to reasonable performance expectations. **25**

CRITICAL PRACTICES

PLATINUM POLICY

Chapter One

Introduction

Let's first identify the basic "raw materials" of product manufacturing—the very essence of the requirements for every product manufacturing operation. There are four primary elements:

Money—for start-up and from profits to prosper
Tools—building, machine, mold, software, etc.
People—and the policy/practices they choose
Product—embodied in design drawings and specs

These four elements must be present in any successful product manufacturing company.

Let's focus again on the last item, the **product—embodied in design drawings and specs. One of the basic raw materials for efficient and affective product manufacturing**.

So why is it a surprise for some to hear that the management of the design documents is a critical discipline? Without precise and controlled design documents, you do not have a producible product. The change process is often identified as the company's most expensive process. The release process is critical for minimizing the time to market. An accurate bill of material is essential to operations folks. The request process is often an irritating "hanging-chad." Without minimum control of design documents via make-sense processes, practices and measurements, you have some degree of chaos.

A certain amount of chaos exists in almost every product manufacturing company because of a lack of basic CM policies, practices, and processes—what this analyst will most often refers to as "standards."

Some folks say that they have attained ISO (or similar) certification so they must be okay in the CM world. Sorry, but ISO certification is only the first step out of chaos. ISO doesn't care if your standards are fast, efficient, measured, or truly effective—only that you have documented them and follow that documentation.

This can be put in perspective by placing your company on the CM evaluation ladder (Figure 1.1).

Since this writer hasn't witnessed a world-class system, the best practical evaluation can only be made by putting together the best of the best practices witnessed. Much needs to be done in most companies to attain best-in-class CM as defined by this writer:

Configuration Management for Senior Managers. http://dx.doi.org/10.1016/B978-0-12-802382-2.00001-5

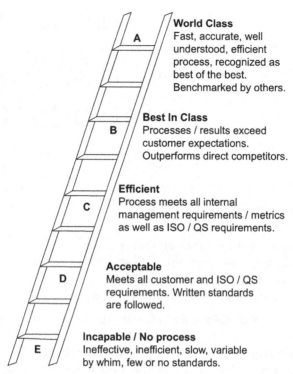

FIGURE 1.1 The CM system evaluation ladder. Adapted from an article in *Quality Process* by DeToro and McCabe.

CM defined: Make sense, documented, fast, accurate, efficient, well understood, minimally controlled, effective, measured, process approach to product design and definition throughout a product's life cycle.

Most of what needs to be done can be accomplished by a proficient CM manager, if he or she has the resources, training, and a dedicated **executive champion**.

Policy: If best-in-class processes are to be attained, every company/ division should have one executive committed to be the CM champion.

The executive champion will be expected to be frequently in touch with, guiding and helping the CM manager through their process improvement or reengineering. That person will remove roadblocks—there are always some in the way of reaching best-in-class processes.

A knowledgeable and effective CM manager can move slowly *toward* best in class by continuous improvement without an executive champion. However, most CM managers will have little chance to continually improve (or reengineer) to best-in-class processes in any reasonable time frame without an executive champion.

Some might say that they can point the way to *world-class* CM. Since this analyst has never witnessed a world-class set of CM processes, it would be presumptuous to claim world-class knowledge. Thus, the writer can only combine the best of the best practices witnessed or devised to hopefully approach a world-class system.

The CM system consists of the following basic processes:

- **Release,**
- **Request changes,**
- **Making changes, and**
- **Bill of material**.

There are two other processes that might also be included under the CM umbrella:

- **Order entry, and**
- **Deviation/waivers/off-specs**.

The field change process will be a separate chapter although it is generally though of as being part of "making changes." This will be done because some companies simply do not have field changes issues and can thus skip that chapter.

If these processes are overcontrolled (as occurs in some environments—especially DOD contractors and large companies) the engineer's creativity is stifled and the processes become slow, costly, confusing, burdensome, and often divisive. If the system is undercontrolled, you will have some degree of chaos.

Many, many details involved in the discipline will be excluded from this work. The writer struggled mightily with the challenge of sorting out the critical information from the 350 pages of *Engineering Documentation Control Handbook* and the 240 pages of *CM Metrics*. Partial success may have been accomplished but certainly some failures as well.

The goal has been to convey the critical, essential information applicable to both military and commercial products.

Executives, especially chief engineers may be wise, if time allows, to read the *EDC Handbook* and *CM Metrics* for the **rules, reasons, and metrics** delineated there.

One chief engineer, after reading the EDC Handbook (on a beach—during a vacation) and then implementing its essence, wrote:

Date: April 21, 2008

Frank Watts
EC3 Corp

Dear Frank,

Inertia Switch would like to thank you for some of the best wisdom regarding configuration management. We were recently awarded several quality awards and received our IS9001 and AS9100 certification last month. A majority of our audit revolved a great deal around design and development control. We received 100% scores in these areas resulting from techniques we used from your book. We are an old but newly managed

aerospace company serving the world. Some customers are NASA, Lockheed, Airbus, Boeing, Raytheon, and Sikorsky. All of these companies have accepted our methods and most have commented that they wish their system was as simple. We feel confident that your "good logic" will continue to grow our company for many years to come. Again many more thanks from all of us.

Best Regards,
Brian DiGirolamo
Vice President, Chief Engineer

Few will take vacation time to read the *EDC Handbook*, but you may find time to read this work. Then get some handbooks and the metrics book for your key people and designate an executive management champion. This will be a giant step toward implementing best-in-class CM.

The emphasis in this book will be to design a set of critical policies, a few essential practices and very significant process decisions that will foster efficient and effective management of your CM world. This in turn, will allow you to set the stage for innovation in engineering and operations while making CM largely transparent.

Good Logic!
Frank B Watts
ec3corp@rkymtnhi.com
www.ecm5tools.com
(970) 887-9460

Chapter Two

Why CM

We have already asserted that control of the design drawings and specifications are essential to the product and that control is called Configuration Management (CM). From a management standpoint, however, what is CM? You have already read the authors definition, but what does it/should it mean in senior management terms?

> **Policy: The product design must be documented and controlled effectively and efficiently for profitable, sustainable, production, service, and sale of our products.**

This consultant, writer, and seminar giver often hears that, "it seems like no two of our products look alike." Knowledge of the configuration of each product would seem to be a logical requirement for any product manufacturer—but more specifically what are the significant reasons to seek best-in-class CM?

THE DIRTY HALF DOZEN

Many reasons for make sense, documented, fast, accurate, efficient, effective, well understood, minimally controlled, measured, process approach to CM are listed in the EDC Handbook. The critical half dozen most important to executive management are as follows:

1. The customer certainly must receive exactly the configuration which they ordered and within the promise to deliver time.
2. Control of the design drawings, specs and bill of materials is critical to any product manufacturer, if repeatable production is to occur and profits accumulate.
3. CM standards and processes can contribute to profitability; by reducing product cost, improving product quality, as well as release and change speed and quality, while eliminating the "throw it over the wall" syndrome and freeing engineers to innovate.
4. Certainly the engineer should want to know what is in each product for trouble-shooting problems in the future. The field support, retrofit, and repair folks also need to know what is in each product for effective customer support. Customers doing their own support functions need to know what is in each product including the service part content.

Configuration Management for Senior Managers. http://dx.doi.org/10.1016/B978-0-12-802382-2.00002-7

5. Changes to design documents and thus the product need to be tracked in order to meet regulating agency requirements and good manufacturing practices.
6. In many companies, the liability issue is critical. Picture yourself on the witness stand when the prosecuting attorney asks; "So these changes were made to correct this critical product safety problem—do you know exactly which units had and didn't have those changes and did you respond quickly to the need to correct the problem?"

Executive management realizes, probably better than most, that speed in the processes is critical to most of these half dozen.

PROCESS SPEED

The speed of the processes is very important. A Harvard Business Review article *Time—The Next Source of Competitive Advantage*, states; **As a strategic weapon, time is the equivalent of money, productivity, quality, even innovation.** Bold claims indeed, but true in this analyst's opinion.

Officers of the company definitely realize that the speed with which a new product is brought on line is critical to market share.

If a change is worth doing, it certainly is worth doing quickly. The change that doesn't need to be processed quickly probably isn't worth doing. This is not to say that all changes should be *implemented* as quickly as possible—they should all be made effective according to the individual need.

THROW IT OVER THE WALL

The operations folks often say that "they" (engineering) "Throw it over The Wall!"—meaning that the new product or changed documents come at them without much forewarning or involvement.

The engineering folks say that "they" (operations) "Aren't around when you need to know how they will process the parts and product"—meaning that they aren't involved early and therefore engineers are being "second guessed."

The engineering folks usually have purchased a PLM system. The manufacturing folks have probably previously bought an ERP system—the systems often don't "talk" to each other. They are frequently maintained by different functions. Two (or more) bills of material are common—my bill, their bill, his bill, should be "our bill."

The engineering folks tend to be cautious and analytical, while the operations folks tend to be shakers, movers, and doers. The engineering people's attitude is generally; "Ready … Aim … Fire!" While the manufacturing people are generally; "Fire/Aim … Fire/Aim … Fire/Aim." Face reality, we are dealing with near opposite mind sets.

The CM organization therefore needs to be the communicator between engineering and the rest of the company—manufacturing operations, field service, QA, materials/supply chain, sales, etc. They need to "Bridge the Gap"

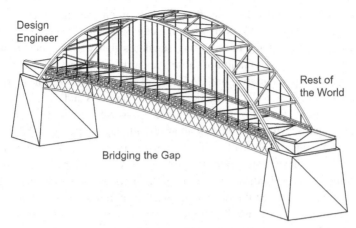

FIGURE 2.1 Why CM is so important.

between engineering and the rest of the company. This is why the discipline is so important—see Figure 2.1.

The CM organization also needs to assure that quality design documents are in the right place at the right time.

> **Policy: CM must be chartered, manned, and expected to be both communicator and the conveyor of new design documents and changes to the right people at the right time.**

This doesn't mean, for example, that CM delivers hard copy or even online copy to the fabricator or assembler. They should know the responsible industrial or manufacturing engineer, and make them aware of the availability of released drawings and specs. CM should also inform the IE/ME of the right time (effective date—with help from production control) to place the revised drawing on the floor.

QUALITY CONTROL FOR ENGINEERING

CM can also logically be expected to be the quality control function for engineering's products—drawing and specifications including code, bills of material, and changes their-to. Since they "handle" every document upon release and every change, they are naturally in a position to assure that drafting and CM standards exist and are met.

The quality organization is probably very well occupied with operations issues and has had little time for documentation issues except to serve on the release or change team/board. In that capacity, they tend to become part of the problem as they are deep in the trees and may fail to see the forest. How can the QA folks be critical of a process when they are intimately involved in it—signing releases or changes? Very difficult! They should seek a quality assurance roll in the CM processes.

A computer manufacturer had five checkers in engineering who routinely "bled" red marks on a proof print when the engineers thought they were ready for release. That company had drafting standards that were in four books of nearly 1000 pages. Examination of those standards and practices revealed that the standards were excessive, pre-CAD, out of date, open to much interpretation, and had not been well communicated to the engineers and designers. Much animosity existed. They directed one checker to oversee rewrite and streamlining the standards, which resulted in about a 100-page document. That person proceeded to educate all involved in the new drafting standards including the CM technicians. They transferred that person to CM with training and standard maintenance responsibilities. They charged the CM technicians to enforce the standards and transferred the other four checkers into design and development work. A windfall of four "new" designers with the animosity eliminated.

We will later discuss the issue of who signs what documents. The quality control role that CM should play will also be noted frequently throughout this text.

CM should also be chartered to measure and report on the processes and to continually improve those processes. The role of the executive champion—a facilitator much needed in the processes—is critical in this regard.

Of course, if you manufacture horseshoes or other inseparable low-tech products you may not need CM—not much anyway.

SUMMARY

CM is a requirement in almost every product manufacturing operation because of the following:

- Design documentation is one of the four critical elements of profitable product manufacturing.
- Design documentation is engineering's product and that documentation is required by almost all the company functions.
- Designs do change and changes need to be accurate, understood, communicated, and tracked.
- Most companies have a degree of whether chaos or overcontrol or both in the CM world.
- A function is needed to bridge the gap often found between engineering and the rest of the company.
- An executive champion is needed for fostering best-in-class CM.

Chapter Three

CM System, Where to Start

The best place to start analyzing the CM system is by measuring the current processes—establish metrics. Without measurement how can we tell if the processes are actually being improved? What are those CM processes, measurements, related executive policies, and the expected performance?

The processes which we need to measure and improve are release, BOM, request for change, and change management (including changing shipped units if applicable). Material rejects/deviations/waivers can certainly fall into the CM world. Other processes such as order entry might well be included under the CM umbrella.

The executive champion should work with the CM and other management to outline the CM turf at your operation and to include measurement of process activities.

PROCESS MEASUREMENTS

Key metrics for all levels of management are necessary in order to assure continuous improvement. Certain metrics should also be established by the CM manager for all the people involved in the processes.

The key metrics for executive management are:

- Process speed—major segments of the processes need to be measured—usually in work days.
- Process volume—the number of releases, requests, and changes completed through each major segment of the process—and the backlog or work in process in each major segment.
- Process quality—a method of measuring the quality of the process output.

Of course there are a number of other measurements that can be added. The CM organization should prepare these metrics and distribute them regularly—at least monthly—weekly if properly manned.

Most measurements will be for the purpose of tracking process performance and improvement. When significant progress is made, the metrics may be viewed as "horn blowing," however, as an old boss once told this writer—a train don't run by the horn, but you never saw a train without a horn.

9

Configuration Management for Senior Managers. http://dx.doi.org/10.1016/B978-0-12-802382-2.00003-9

The executive champion should consult with other senior management to assure that the metrics chosen are most meaningful for the current conditions. Executive critical metrics will be suggested in this book. Many other likely metrics may be found in *CM metrics* as well as benchmarks, survey results, and CM principles.

DISTRIBUTED CM

One of the mistakes frequently made in larger or DOD contract companies is to create a CM organization for each project office. Yes, there are occasional differences between CM requirements for various projects/customers/agencies that may have caused this condition. Yes, it is nice from the project perspective for each project to control their project CM. However, this analyst has never seen distributed CM work well. Distributed CM opens the door to chaos or overcontrol or both.

The overwhelming majority of the customer/agency requirements can best be met by a single set of processes with the ability to tailor the process to occasional unique customer requirements.

Policy: One CM office shall serve all the projects in one logical business unit.

Of course the decision as to what a "logical business unit" is open to considerable interpretation. Generally, it would be a business unit that includes engineering, operations, supply chain, and product support.

The single CM organization can, and should, have the CM technician(s) physically sit with the project people.

The extreme of distributed CM is to have each engineer or project engineer responsible for their own CM. Some examples of the chaos that occurs with distributed CM are as follows:

Example: The engineer/project office decides that it is expeditious to work directly with a supplier and give them design and development drawings for prototype parts without the buyer knowledge.

Result:
1. The best supplier for production purposes might not be chosen.
2. The best price for pilot or production units might not be obtained because the buyer will now be "negotiating from jail."

Example: The purchasing or fabrication folks need the part drawing now because it is a long lead item to satisfy the company production schedule. The engineer/project office decides that it is okay to give the drawing to the buyer or supplier even though it *hasn't* been reviewed by the team and properly vetted.

Result: Oops—a quarter of a million dollars worth of parts show up on the dock and are found to be unusable.

These are not unusual scenarios. This analyst has witnessed exactly those results and seminar attendees often report the same outcomes.

All product manufacturing companies/divisions need *a* person/function to perform the CM function. The alternative is to allow each engineer, designer, engineering manager, or project office to do CM as they think proper—a formula for chaos.

At one company, the supply chain folks were deciding which changes to transmit to the supplier and which not to transmit based upon which project office the changes came from. They considered the revision level changes from one office to be "insignificant." That project office function had not involved the supply chain folks in the change process and was ignored for all revision level changes—a laughable but actual situation. Combining distributed CM into one divisional function with one set of practices (with exceptions documented) solved that problem and they ultimately required fewer CM techs.

Every engineer or project left to be responsible for their own CM not only results in chaos (as many processes as there are engineers or projects), but it also fosters a climate where it is all too easy to violate basic principles for effective release or change—with good intent.

ORGANIZATION

Do you have a CM function—a person or persons designated to be in control of design documents after release? Who should they answer to? How many people are required?

Policy: One CM "organization," usually in Engineering, needs to be established—part of a person in a start-up, a person as you grow, and several people when successful.

Policy: When your operations are multiplant, a slim corporate function with each plant/division having a CM function is appropriate.

This presumes that each division has both engineering and manufacturing operations.

The corporate CM person would specify certain critical requirements that all divisions must follow, but allow the divisions to innovate on the remainder of the processes.

The slim corporate function is needed to ensure that minimum standards are met. This minimum level of standardization should be focused on three criteria:

1. Moving a product from one business unit to another.
2. Customers contact (or contract) with more than one division or business unit, documentation should look alike to the customer.
3. Field service is done by a single person for products made in more than one division.

For example, the part and document numbering system and interchangeability rules should be identical in all the divisions to satisfy all three of the above criteria.

The corporate CM person would also specify a few metrics to be used by all divisions. They should also cherry-pick successful division process elements and communicate those elements to other divisions.

Very large companies might well have an executive champion at the corporate level as well as one in each division. At the corporate level, the slim CM function should probably answer to the corporate executive VP of engineering. The divisional CM functions should also normally answer to the division engineering functions.

CHAIN OF COMMAND

This analyst has seen the CM function perform well and poorly when part of engineering, quality or operations. However, the CM function should normally answer to the engineering organization. See Figure 3.1 for survey reporting results.

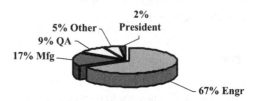

CM Organization Answers To

FIGURE 3.1 Organizational responsibility.

"But isn't that like having the fox watch the chicken coop" some ask. Yes it is, but they (drawings, specs, code, bills of material, and changes thereto) are engineering's chickens! If engineering eats too many chickens, then move the organization elsewhere or change the management.

> **Policy: The CM organization should answer directly to the Chief Engineer or to the Director of Engineering Services—not any lower in the organizational chart.**

A CM function buried in the organization will lack the clout required to achieve best-in-class processes. If CM is in operations or quality, they should be at a similar level in the organization.

CM MANAGER

Much can be found about the CM managers responsibilities in the *EDC Handbook*. A good job description is included in the writer's standards manual. Above all, he or she must be a bridge builder—to bridge the gap between engineering and the rest of the organization. Since the function is controversial (probably right behind wage and salary), the manager will perhaps never be *liked*, but they can be *respected* by adhering to the principals in this book. Their training is critical and should be ongoing.

ISO STANDARDS AND CERTIFICATION

The vast majority of ISO requirements are CM requirements. Companies that are certified have written standards and follow them to a degree accepted by an ISO certifier—however:

> **Policy: ISO certification must be recognized as only the first step out of chaos. Standards implementing the best of the best practices should follow.**

It is worth repeating—ISO doesn't care if your standards are fast, efficient, measured, or truly effective—only that you have documented them and follow that documentation. ISO would, for example, accept distributed CM (each engineer/project office doing its own CM). They would also not care how fast or slow the processes are, how efficient or effective they are, or even if the process speed is measured.

The ISO certifiers generally like the standards to be structured as the ISO standards are written/paragraph numbered. Many companies have done this and found that the ISO rewrites the standards on occasion and thus their standards have to be renumbered to comply—better to organize standards by process.

Never accept that certifier's word as the gospel. Read the ISO standards yourself and do not hesitate to challenge a certifier if and when it seems appropriate. The most famous example concerns the "*latest* rev print at the point of use" requirement that was in the early ISO standards. This statement didn't recognize that some changes occur at a future date. Thus the ISO spec was finally changed to allow the *proper* rev level print to be at the point of use.

If your company already has ISO certification, you can begin to improve your processes from that base. If not, don't plan to get certification until you have completed a process improvement project(s). When you have completed process improvement on all CM processes, ask yourself if you should now get ISO certification... or not.

There is a growing tendency for companies who don't do business in the European Union to ignore ISO all together. Unless you do business in Europe or have some compelling interest in certification, consider self-certification by developing good metrics and constantly improving performance.

It is all too easy to consider the CM world to be in fine condition because certification has been achieved. Never forget, ISO certification is only the first step out of chaos. Also never forget that ISO certification is expensive and must be renewed frequently.

DOD/MILITARY STANDARDS

The military/DOD folks would also prefer that the standards be organized as their specs have been written:

- Planning
- Identification

- Control
- Status Accounting
- Audits

Companies who do business directly with the DOD/military are prone to use this method of organizing their standards. Because of this influence, much of the discussion about CM in the Internet is fraught with those terms.

Most military contractors begin by hiring a CM manager with military contracting experience. That person gets out all the DOD and branch specifications and designs a CM system around those specs. Many of the DOD specs are confusing and complicated—a major reason for the $400 hammer.

A much better approach is to design a CM system that satisfies good commercial practices first and then get out the Mil Specs applicable to your business and add to or modify the system to fit those standards.

Policy: Satisfy good commercial CM practices seeking the best of the best processes, then look at modifications to satisfy unique Mil Spec requirements.

As an example, take a look at the specs involving the choice of CI—Mil Spec Configuration Item. Look at the *LinkedIn* discussions about CIs. The discussions, debate, and interpretations are absolutely overwhelming.

Now ask yourself, to satisfy good commercial practices, what assemblies in your product would you serial number and nameplate—thus tracing certain changes to that item? Having answered this question you have very likely determined which items are your CIs. Submit that to your contracting agency and modify if required.

Note: This writer will, in this text, often use the term serial number to imply any form of **individual unit identification**—date code, batch number, order number, mod number, etc.

Keep in mind that customers have been known to take items out of product(s) that are out of warrantee and put them into a product that is in warrantee and return that product for warrantee repair—cannibalization it is called.

Realize that good commercial practices require that a file be kept of SN(s) into SN—specifically to use in warrantee claims. Some Mil Spec experts will cringe at this simplistic thought but could it be that they have a vested interest in keeping the discussion as complicated as possible?

Most military subcontractors will find that good commercial practices will satisfy all DOD requirements—see the earlier quoted letter from the *Inertia Switch's* VP and chief engineer.

Even first-tier military contractors should start by developing best-in-class commercial practices and then examining the Mil requirements for unique refinements.

BEST OF THE BEST PRACTICES

In this analyst's opinion, neither ISO nor DOD standards are the best way to view, learn, teach, apply, or standardize your CM policy and procedure.

It is far better to organize by the **processes** that are involved in every product manufacturing company. As Morris and Brandon wrote "**Processes Are the Essence of Business**." The major CM processes are:

- **Release—of part, assembly, code, product, or document**
- **Bill of Materials—both ERP and PLM**
- **Change request or engineering action request**
- **Change Management**

Other ancillary processes such as order entry and discrepant material/deviations should normally be included in the CM arena.

Of course a "general" category is also needed in order to cover subjects common to all the processes. This writer has written a set of generic standards, including flow diagrams, which will give your organization an excellent start at standards development.

As you will see, each process will require certain standards to be written in order to clarify the requirements and work flow.

Again, the best of the best practices will suit both commercial and military organizations with few modifications for agency requirements.

STANDARDS RESPONSIBILITY

The CM function needs to be chartered to assure that make sense, documented, fast, accurate, efficient, well understood, minimally controlled, effective, and measured processes are put in place by written standards.

> **Policy: The first "best-in-class" general standard written should outline the responsibilities of the CM organization.**

CM should play a key role in attaining the lowest total product life cycle cost—they should therefore be responsible for all design document release, change request, design change, bill of material, and related process design and control—such as the order entry, discrepant material/deviation processes.

The CM manager should obtain the signature of the executive most affected by each standard—after a through review and comment process by all affected parties. The executive champion should assure that only one executive signature is required.

It must be recognized that standards unwritten or not followed will produce a certain amount of **chaos.** If a standard is not effective and efficient, it should be revised immediately.

Each standard must indicate who is responsible for keeping it up to date. Each should also note how exceptions are made and documented.

Senior management needs to take part in the development and enforcement of the standards.

Critical practice: The senior management needs to "get in the face" of any engineer, manager, or other person who violates the CM standards.

It must be made quite clear that a certain minimal amount of control is a top management priority. Eye-to-eye, face-to-face is best. This will probably only have to happen once or twice, because the word will quickly spread.

PEOPLE REQUIRED

Most companies have a document control function. That document control function is often manned by one or a few low-paid people who are ill trained, buried in the organization structure, frustrated, and ready to change jobs.

A CM function that is properly managed, organized, trained, and manned can tear down the wall and bridge the gap between engineering and manufacturing. "Properly manned," however, does not always mean hiring new people. Often the people are there, but they are just scattered in other parts of the organization.

For example:

- The people who input **design data** to the ERP BOM are often in operations—but should be in CM.
- The designer/draftsmen who incorporate design changes into the master documents are often in the Design Drafting function—but should be in CM.
- The people in project offices doing independent CM should be in a singular CM organization.

The number of people required for a CM organization will vary with the responsibilities, customers/agencies, and the size of the company. As a benchmark, the following data may help. It is from the author's comprehensive EDC/CM benchmarking **survey** of 58 companies/divisions producing a variety of products.

Realize that few of the companies surveyed included change designers/drafters in the CM function. Few had the ERP BOM design data entered by CM. Few had embedded software. Also realize that few did any significant measurements/metrics on the CM processes and that almost all produced nonmilitary products.

The numbers should therefore be viewed as bare minimum manning.

People in Company	Average People in CM
0–100	2.2
101–500	6.0
501–1000	8.0
Over 1000	12.3

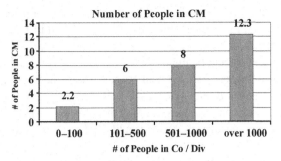

FIGURE 3.2 CM minimum manning.

In bar chart form—see Figure 3.2.

The manager was included if they were full time. The range of CM employees in all 58 companies surveyed was 0–35.

DOD functions require more man power than a purely commercial environment. So would inclusion of change drafting and ERP/PLM BOM data entry—both of which this analyst strongly recommends.

FUNCTION NAME

The title you give/have given to the function is not particularly important. Configuration Management is often used and is most expressive currently. However, a Google Search will show many more software CM hits than product CM hits so another name may be appropriate.

The CM term also has heavy DOD implications. Thus if you are in the DOD/military business, the CM name is very appropriate. If you are in the commercial market then another name may be appropriate—Engineering Documentation Control, for example. Since we engineers hate the word "control," perhaps Engineering Document Management.

The CM organization might well be combined with the Design Drafting and other functions as part of an **engineering services** organization. The name you give your CM function isn't critical but the responsibilities placed/not placed in CM are critical to the outcome.

RESPONSIBILITIES

This writer divides the CM responsibilities into two parts—the traditional document control and CM.

Document Control Function Responsibilities

- Assign all part numbers, request numbers, and change numbers for all design documents.
- Control each master design document after its appropriate point of initial release and assign document revision levels.

- Input and accuracy of the BOM database(s) (design information/data) regardless of the software tools involved (ERP and PLM).
- Change request monitoring.
- Change control and facilitation.
- Chair the release, request and change team meetings (often called Change Control Board—CCB).
- Backup document databases or assure that information technology backup practices are satisfactory. An off-site file updated at least daily is a must.

Note the responsibility for assignment of revision levels to design documents after the appropriate point of release. Any process that allows an engineer to assign the next revision level to released design documents creates a truly uncontrolled environment:

Example: The engineer forgets to change the rev level on his design document and the supplier builds to an older print.

Result: Still more bad parts arrive on the dock—or added supplier cost/waste that you end up paying for.

This and other such conditions call for a critical policy:

Policy: CM must have control of rev levels after release, or you do not have control.

After the appropriate point of release, the assignment of the next revision level must be solely in the hands of CM. It should be done only after review and approval of the documented change is complete. **The assignment of the next revision level on affected documents is thus an indicator that a release or change has been properly reviewed and approved and that all parties affected can proceed to implement the change with very low risk.**

The input of design data to the ERP system is frequently not included in the document control/CM responsibilities, but should be. Just because manufacturing purchased the ERP, it shouldn't have the responsibility for the input of design data (PN, name, description, unit of measure, etc.) and changes thereto. Likewise, CM should do the initial input and changes to the PLM system.

Start-up companies sometimes have a one-person document control and CM function. They often control manufacturing, quality, field support, and even sales technical documents. As the company grows, however, control should move to the organization that authors the documents. But one might think: "That will make several document control functions." Yes, it will, but it also places the responsibility for control with the organization that authored the documents.

Of course, the distributed document control functions must be "tied together." This is best done with work flow diagrams and other CM standards which are audited and controlled. The distributed document control functions need to be coordinated. Such "umbrella responsibility" belongs with the CM function.

CM Responsibilities

- Coordinate all the technical document control function activities.
- Train all key personnel on the basics of CM and the company CM system.
- Ensure that nonvalue-added requests for change are filtered out (the primary purpose of the request process).
- Develop metrics—continually measure and report on the CM processes as to speed, volume, backlog, and quality—at a minimum.
- Do the revision (incorporation) drafting for design document changes.
- Control *all* design data and document transmittals to customers and agencies.
- Ensure that the changes are tracked to the actual date or specific units affected (via serial number, for example). Sometimes called change effectivity tracking—the DOD term is status accounting.
- Ensure easy access to a tracking database and assure that reports can be produced as required.
- Ensure the quality of all design documents and changes thereto via application of the CM and drafting standards.
- Audit the system to ensure that it is followed or changed when appropriate.
- Benchmark the CM system and ensure continuous improvement of the CM processes.

If any of these functions are not included in the CM manager's responsibilities, the results will likewise be limited.

REENGINEER OR CONTINUOUS IMPROVEMENT

Whether to "bootstrap" improvements or redesign the CM process is a significant and difficult issue. This analyst would suggest that the choice depends upon whether or not the culture and the top management are ready for redesign. If they are, and a dedicated, inspired executive champion is involved, then reengineering is doable. Without a culture ready for change and such a person involved, incremental process improvement is the practical choice.

Chances are that your CM manager is already trying to do some process improvements—bootstrapping betterment. The executive champion needs to work with the CM manager to figure out what approach to take. Lacking an executive champion the CM manager has little choice but to continue making incremental improvements.

The executive champion may choose to have the CM manager continuously improving the CM processes and to give him or her support to accomplish this.

Policy: Any process designed by man can be improved by man.

Even after reengineering a process, follow up with continuous improvement.

The change process is often experiencing the most issues. Engineers are very reluctant to release a new design in lead time when they know that the change process is slow and painful. Thus the change process may be the place to start. However, most of the general standards we mentioned must be in place first—see Chapter 5.

The defective material/deviation process might also be a place to start because it is often problematic—as is the order entry process. There is no magic formula for where to start. One company started with the field change process because bad rules were being used to decide which changes to retrofit and were thus wasting money by the bag full.

This analyst has sometimes recommended that the defective material/ deviation process is the place to start because it was being abused by making it a way to make fast (uncontrolled) changes. However, this can only be done if there is a reasonably fast change process to rely on.

PLAN TO IMPROVE

Reengineering any of the major processes is a daunting task and should probably involve some consultation. However, the VP/chief engineer quoted earlier did it with a copy of *Engineering Documentation Control Handbook*.

No matter what process is chosen for improvement, some overlap will be apparent and it will be necessary to find ways to avoid taking on more than one process at a time. Start by establishing key metrics for the chosen process.

If reengineering, address one process at a time and flow diagram the *current* condition. Assure that this current flow diagram is agreed upon by key parties. At a medical device company, this analyst witnessed current condition diagrams for the same process made by engineering, IE, and CM that described significantly different current states.

Put a small working task team together—preferably one person from engineering, one from CM, and one from operations—perhaps an additional person from service in some companies. **A team much larger than three or four people is bound to fail** in this analyst's experience. This rule was learned during the author's outhouse tipping experiences—too many people involved will result in someone falling in the hole—not a good outcome!

Propose a new work flow diagram that streamlines the process. Design forms and form instructions for that process. Write the associated standards.

The task team would develop standards as we have discussed. Drafts must be circulated, revised, recirculated if necessary before signing and implementation.

It may or may not help to have a larger "steering committee" involved to add some perspective and/or to facilitate "buy-in" from multiple people/functions… or to muddy the water. A high-level, cross-functional committee meeting once a month or so might be wise depending upon the company culture. The executive champion must carefully consider the methods of operation of the task team and steering committee. They must not be process designers but rather be there for understanding the logic used by the process design team.

When the new process has been thoroughly vetted and signed-off, it should be implemented—done on one product line at a time or across the board—another tough question for the CM manager and the executive champion.

Without an executive champion, the CM manager should implement changes in small steps—with or without a small team—bootstrap continuous improvement—essentially one standard or process change at a time. Take small bites and chew them well.

See the *EDC Handbook* for much more detail on implementing a new process.

SUMMARY

To get started, get the horse in front of the cart.

- Put a singular CM organization in place with clout, a bridge-building CM manager, and an inspired executive champion.
- Assure that CM has the required responsibilities and manning to accomplish them.
- Get critical process measurements/metrics started.
- Define general standards that need to be written/approved and assure that CM is manned, develop general and process standards, and to implement them.
- Make a plan to continuously improve or reengineer the CM processes one at a time. Small team desirable.
- Set aside Mil, ISO, and similar standards until best commercial practices are completed.
- Modify best-in-class processes to suit the agencies.
- Continue to continuously improve.

Chapter Four

CM and Software

Let's first distinguish between software apps for internal company applications (tools in the design and CM processes) and those which are embedded in the product.

- Process software = PLM, ERP, CAD, etc.
- Embedded software = Engineering software designs which are necessary to make the product function as advertised.

There is a black and white difference in the role of CM in each category which will be discussed in this chapter and in others to follow.

First, it is necessary to disclose some bias on the part of this writer about process software applications in general.

SOFTWARE'S CURRENT CONDITION

This writer just finished reading a 170 page book about SCM information technology. The words "Digitized Platform" were used at least 250 times in that book—but the term was never defined! I searched the Internet for the term and it gave me little help. Guess you would have to hire the authors to find out what it means.

There is a painful lack of logic, use of the King's English, and intuitive design; in the literature, on the Internet, and in the design of many process software apps. Somehow the IT specialists have managed to frequently forget that some "human engineering" is required. For example:

- Don't welcome me to my own computer, I thought that I owned it and the software? Thank me for buying your stuff!
- State messages in good English! Get an English major or good tech writer to help.
- Don't mess with my software without my permission and without clearly stated reasons.
- Quit lying to me—such as telling me you are updating my software (*installing updates—one of one*)—almost never more than one, *every time* I sign off.
- Just save my data every 5 min so I don't have to.
- Few CM, PLM, ERP process software apps can be installed properly without significant supplier involvement—often more costly than the initial purchase of the app.

Configuration Management for Senior Managers. http://dx.doi.org/10.1016/B978-0-12-802382-2.00004-0

- The author has owned his desktop computers with operating systems for years—why in the world don't you have it right by now? Then you will quit "supporting it"—before ever getting it right.
- And when you quit "supporting" it, the result is that it works better than before.

Yes, I know that hackers are responsible for many weird happenings—but why not tell us that—and go after the hackers?

When discussing CM process apps at seminars, there will frequently be pairs of attendees—one buying a particular application and another dumping the same app. One can often blame the implementation in the company dropping the app, but the software supplier isn't without considerable blame—confusing terminology, unclear literature, unclear choices, bugs, etc.

Sometimes this author often thinks that **IT** stands for **I**diots **T**inkering. To paraphrase Mark Twain; "**If the software developer's funeral were to occur tomorrow, I'd postpone all other recreation to attend!**" Joking of course, but you get the drift. Now that you know this writer's bias, let's push on!

MODERN TREND

There is a regrettable modern trend to seek a software solution for every process problem, often when the "problem" hasn't been carefully defined—app mania!

Critical Practice: Perceived problems in the CM processes are often addressed by buying another software app.

Result: Best outcome—bad processes executed faster and no fundamental improvement in the processes. Worst outcome—chaos!

Attempting to solve process problems with software application solutions is a touch of insanity—repeating the same behavior over and over, but faster, and expecting different results—especially when the "problem" often has not been well defined.

Policy: CM processes with current software shall be addressed first, measured, and brought to a reasonably fast, well understood, efficient, effective, minimally controlled state, and then automated for efficiency if necessary.

This will result in efficient and effective processes.

PROCESS REDESIGN BEFORE SOFTWARE

Morris and Brandon in *Reengineering Your Business* said; "To be sure, information technology was used to support the new process, but the process redesign came first and the technology considerations second."

Hammer and Champy in *Reengineering the Corporation* said; "**Some people think that automation is the answer** to business problems. True, computers can speed work up, and in the past 40 years business have spent billions of dollars to automate tasks that people once did by hand. **Automating does get some jobs done faster. But fundamentally the same jobs are being done and that means no fundamental improvements in performance.**"

Thus, to attain fundamental improvements:

Policy: Put any proposals for purchase of new—or major changes to— ERP, CM, PDM, PLM software on hold until the CM processes all measure up to reasonable performance expectations.

If minor changes to current software applications are suggested by the process improvement team, they might be done with that project—certainly with metrics in place to measure improvement.

Listen to the IT folk's and others objections, but hold fast. See that the problem is properly defined, that the processes are measured and that the processes are improved to reasonable expectations with legacy software and current manual elements before adding or modifying existing software.

Fortunately, the embedded software engineers are often of a different ilk.

Where the product CM organization enters the picture is worthy of definition.

CM'S ROLL IN SOFTWARE

For process software—ERP, PLM, CAD, etc.—CM should be part of the selection and implementation team. The elements they should look for/work toward are in the *EDC Hand*book.

For embedded software, the CM roll should be very similar if not identical to the roll they would have in any CAD produced or hand-drafted drawing or spec.

WHAT CM SHOULD CONTROL—AND NOT

This book is not about controlling the ERP, PDM, CAD process software or any homegrown process software tool. It will be about the *use* of such software apps, especially with regard to the bill of material.

This book is about minimum control of the releases, requests, and changes for embedded *product* software by the CM organization just as they will control the mechanical, hydraulic, or electronic designs.

The embedded software need not be controlled by the CM organization in any way during the engineer's design and development role. There is no need of control (by the product CM organization) until release for pilot production— just as with CAD mechanical, electrical, and hydraulic designs. This allows the software engineer the freedom to innovate before release, without the control necessary for operations, service, and customers.

The CM organizations *do* need to control any transmission to a customer and all releases—both initial and changes.

SOFTWARE (SW) AND FIRMWARE (FW) CONTROL

Product SW and FW are essentially identical issues from a product CM perspective. The code evolution of software design and development are controlled by the software engineer—when the software is in the design and development phase. Subsequent changes/releases remain in the design and development phase, under the software engineer's control, for each version until ready for release.

> **Policy: What goes on inside software engineering is essentially R&D design phase business and need not be tracked by CM except for tracking Requests for change and customer submissions.**

For a discussion of requests, see that chapter. The software engineers should keep track of all changes to the code while in design—with or without an SCM program. The entire code must be tested together (before release or change) as required by the IT and senior management.

In the initial release of the SW/FW, a specification document is needed to identify PN, name, and initial release/version number. This spec document must describe the media which accompanies the document on release. This document should have the following:

- A standard part number.
- A description of the media.
- The applicable product(s) used-on.
- Two copies of the media marked with PN, name, and version—Example: PN 12345601, Indexing Functions and Version Number 1.0.
- Describe the location and version of the source code, build environment, tools, settings, and other data pertinent to reproduction of the code and media.

The last element might well be done by furnishing a zip file archive to CM.

When CM verifies that all the needed data are present they will assign the initial rev and enter the PN, rev, etc., into the PLM and ERP.

Subsequent changes to the code should be done, tested, and released via change order and reflected in the revised specification document. When CM finds all the needed data present, they should:

- Roll the tab (part number) of the document since all released changes to software are, by definition, noninterchangeable.
- Assign the next Rev to the document.
- Make the necessary changes to the ERP and PLM.
- Forward the media to production with a copy of the specification.
- File the backup copy

Software release change orders need not describe the software changes in that release precisely but they should always list each request that *has been satisfied* with that software release. CM should make sure that this list matches the request status log.

It is also advisable to list in the software change order, the *known requests that have **not** been solved in the current release*. This positive–negative approach tends to bridge the gap between software engineers and the rest of the company.

SUMMARY

For product embedded software:

- Treat the development of software as much as you would treat the development of hardware.
- CM controls the software releases.

For process software:

- Avoid App Mania.
- Get processes working well with current software and then consider additions or modifications.

Chapter Five

Standard Foundation Blocks

Before attacking improvement or reengineering of any CM processes, it is wise to carefully put certain general issues to bed. The best way to do that is to develop a separate (from the *QA Manual*) *EDC/CM/PLM Standards Manual*. Why separate—because different review/approvals are usually required and a pride of ownership is desirable for the CM folks.

STANDARDS WRITING

This analyst, when doing consulting "needs analysis," is sometimes presented with a 70- or 80-page document that "defines" the CM system. Besides the CM manager, this analyst may be the only person to read and try to understand the requirements. Better to take small bites and chew them well.

Short standards on specific issues need to be developed by the CM manager. The executive champion can be of significant help in setting policy, developing, and obtaining approval of these standards.

Standards must be on **single subject** and short—one to three pages. If they run to more than a few pages, there is probably more than one subject involved. They should be approved by the VP/director most affected and then followed by training and implementation.

> **Policy: One subject, one standard, one approver, followed by adequate training.**

A lady CM manager at a high-tech tool manufacturing company had a good mark-up/redline standard. It was two pages long and is accompanied by examples of an acceptable markup of each type of document they used. The examples were on the wall in the CM office. She had trained all the existing engineers, designers, and others who might originate a change. Part of every new engineer and designer's training was for her to give them a copy of the standard and walk them by the wall—announcing—"Here…you do it like this and they will pass easily, if you don't, they will come back to you!"

Details about the general subjects which need to be covered are found in the *EDC Handbook* and a generic standards manual is available from the author.

Configuration Management for Senior Managers. http://dx.doi.org/10.1016/B978-0-12-802382-2.00005-2

Each standard must be reviewed by those involved in that specific subject, rewritten and rereviewed as needed and presented to the executive champion for review. The champion should then designate the proper approver—engineering services director, chief engineer or possibly the COO or president.

Each individual process must also have standards (including work flow diagram) supporting the process. Those will be addressed in each applicable chapter.

GENERAL STANDARDS

Some standards do not "fit" comfortably in any of the CM processes (release, request, change, BOM) or are common to all or several of them. Some of these standards would be corporate and some divisional. See the *EDC Handbook* for more discussion. The more important of them are, in no particular order:

Company/Division Policy Statement—Outlines the CM "turf" and general responsibilities.

Standard on Writing Standards—To obtain short, easy to read, well-understood process documentation.

Product Specifications—Includes the content and format of critical product criteria that will be committed to the customer. This standard will require that the product spec be released one phase ahead of all other design documents.

Drafting Standards—Includes allowable drawing and specification formats. Also covers EDC/CM requirements for content of design documents.

Doc Groups—Lists all company technical documents separated by the responsible function and defines basic control responsibilities—IE/ME, quality, sales, tech support, etc. The design documents will be a separate group under CM control.

Cognizant (responsible) Engineers—Defines the responsibilities of design engineers regarding the design and its documentation, test and its documentation, sales and its documentation, etc.

Part Numbers—Defines the attributes of the part number, document number (hopefully not separate from but embedded in the part number) and specifies company policy on tabulating documents and assignment of part numbers.

Interchangeability—Defines interchangeable, noninterchangeable, compatible, and related policy.

Part Number and Revision Level Changes—Specifies when to change part numbers and revision level.

Approved Manufacturers List—Controls the acceptable manufacturers of a purchased item. May or may not include the acceptable suppliers.

Teams—Defines the makeup and responsibilities of teams for every CM process.

Signatures—Specifies what functions should *review* and which should *sign* design specs, design code, drawings, CM documents, and red lines.

Prints, Point of Use, Paper-less—Defines the company policy regarding distribution and control of design documents.

Class Coding/Naming Conventions/Group Technology—Specifies the company policy and procedure to be used.

Nameplate and Serial Number—Defines the requirements for the nameplate drawing and serial number (or lot number, date code, etc.) assignment.

Acronyms and Definitions Standard—Provides a single location for approved definitions and abbreviations in order to avoid placing definitions in every standard that uses a word or term.

The most critical of those and their most critical elements follow in no particular order.

ACRONYMS AND DEFINITIONS

The folks involved in the CM processes have come from different backgrounds, companies, education, and therefore have different terminology. Those differences can best be bridged by an Acronyms and Definitions Standard.

At one client, this analyst sat in a meeting that was attempting to agree on the phases of new product release. After more than an hour of apparent disagreement, we defined terms like pre–pro, pilot, sample, alpha units, test units, etc., and afterward found considerable progress and agreement.

This standard will negate the need for putting definitions in each standard and the risk of having varying definitions.

The CM manager who is attempting process improvement or redesign should write this standard. Folks will more rapidly improve their communications with it in writing—properly implemented with training.

This standard might also include the data dictionary—specific information about each data element—for example,

- Official source of the data (example: PN source is the title block of the drawing or specification)
- Who enters the data (hopefully done only once—PN entered by CM upon assignment)
- Character definition (example: PN=NNNNN-NN where N is numeric, the dash is not to be data-entered and last two digits are for tabulation of the item depicted)
- English definition

Sounds like a no-brainer but you might be surprised by the number of companies who are operating without such a standard and are suffering to some degree as a result. One government site seminar attendee reported that they had software systems with 16 different definitions of a part number—take your pick!

One element that is easy to define, but very hard to optimize is the part number/document number. Several issues need to be addressed.

PART NUMBERS—SIGNIFICANT OR NOT

There are zealots on both sides of this issue. Most product manufacturing enterprises, at one time or another have this controversy. Every new start-up company has this issue. The issue permeates many operations. It crosses all functional lines.

Some say that significance (smart number) is the only way to go. Others say that nonsignificance (dumb number) is the only way to go. In this writer's opinion, either extreme is generally wrong.

A generalization for items from the top level to the bottom level of your structure—for every company—cannot be made to stand the test of logic. Also, any discussion of part number design without inclusion of interchangeability considerations is ludicrous.

While believing in minimum significance, this writer would submit that at the top level of your product, the smart number might make very good sense. Consider the company wherein the salesman or customer structures a smart number from a catalog when determining what features and options the customer wishes (catalog configured). This might be done on line or from a hard copy catalog. Should we take the order by catalog number and have order entry convert it into a factory part number? Won't the conversion inject a possibility of error into the process? When the conversion error occurs (and it will), the customer will receive something different than they wanted—an intolerable situation. In this circumstance, why not use the configured catalog number as the top level/end item part number in the system. If maintaining a catalog works for you, look seriously at using that smart catalog number as your end item number.

The same logic might apply to other "configurable modules" within the product that are separate line items on your sales order.

If your catalog is too difficult to maintain because of numerous/added/changing features, then a minimum significant number for the top level might be in order. A modular bill of material or configurator module at the top level could be the answer.

If you have a smart number and it is beginning to breakdown, or you are merging several operations into one, a good rule to follow would be the following:

Policy: Put very little significance into the part number with the exception of a tab/dash suffix (to be discussed shortly).

Significant numbering systems tend to breakdown. No matter how good you are at anticipating the number of digits you will set aside for a given characteristic, at some point it won't be enough. There are many pros and cons to each practice (significant and nonsignificant) but this potential breakdown is the straw that prompts minimal significance.

See the *EDC Handbook* for a discussion of all the pros and cons.

CLASS CODING

The temptation to use a significant part number throughout the structure is high. But before exploring significance in lower-level item numbering, let's ask ourselves "why significance at all?" The significant part number helps us to find similar parts. If we don't have significance in the part number, how do we search to find similar items? How do we avoid reinventing the wheel? How do we find a temporary substitute for a part shortage? How do we know which parts might be manufactured in a single cell? Or perhaps your history has produced several part numbers for interchangeable items and you want to standardize.

A group technology or class code system (smart number) *outside* the part number may be the best answer. With the power of computer word searches, an intelligent description field or "naming convention" (another form of class coding) might be the answer. With the advent of low-cost computing, it is probably better to set up a database with a separate field for each characteristic that might have otherwise been put into a significant part number. Then searches can easily be done on those fields.

If a company has a couple digits of class code in their current part number, although undesirable for most, it might be serving them quite well. It thus shouldn't be tampered with. If you have a class code prefix of a couple of digits of alpha or numeric it might be acceptable to keep it, especially if that allows you to "sell" the new number format. Thus the number would be the following:

XXYYYYYZZ—where "XX" is the legacy class code.

It could also allow the base number to stay the same for a printed circuit board, artwork, and assembly—a minor aid for the electrical types.

The "ZZ" is for tabulation of similar items on the same document and for noninterchangeable changes to the item. Repeat—for noninterchangeable changes to the item—a version number if you will.

INTERCHANGEABILITY AND THE PART NUMBER

Hours and hours of time that could be spent innovating are spent debating interchangeability and part number change issues. Good interchangeability/PN change standards will not eliminate the debate but will significantly limit it.

Service and manufacturing people end up with bone piles of material because of the confusion created. Failure to change part numbers upon noninterchangeable change sometimes leads to a decision to "upgrade to the latest revision level"—a hugely expensive decision. Is retrofitting a cost reduction? Is installing a product improvement in all products cost-effective or wise?

The question must be asked: "what if we change an item noninterchangeably after release?" Answer—the part number should change! Ever since Eli Whitney, noninterchangeable changes have meant a different part number—**in one or more characters.**

Some operations say "We don't change part numbers—only document revisions!" This is a very expensive policy. It works okay when the product is young and almost all changes are required to meet specifications—and repair and retrofit will be done with the latest rev level item. But what about the mature product wherein most changes are made to reduce cost? You certainly don't want to retrofit cost reductions or confuse your service folks, so we better find the best way to accomplish part number changes for noninterchangeable changes.

The answer is to attach a "tab" or "dash" number on the part/document number. Thus upon a noninterchangeable change, the number would change from tab "01" to "02" and a new document is not needed. It also allows the document revision level to be set aside for document only and interchangeable changes.

> **Policy: Always tabulate the part number. If you have an existing PN convention that does not include this feature, add it as soon as practical.**

Thus, with this **mostly nonsignificant** part number, we can:

- Identify noninterchangeable parts uniquely.
- Embed the document number in the part number and thus avoid cross references.
- Document similar items on the same drawing.
- Save a little labor to prepare separate documents and to revise them.
- Be friendly to those many people who memorize part numbers.

- Take away an excuse for not changing part numbers on noninterchangeable changes.
- Have a "version number" (the tab) for repair and refurbishment people.

This ideal part number **is significant** by virtue of embedding the document number and having a tab suffix.

As stated before, the significant part number may also have proper use at the top level (end item or configuration item) in some companies. It is therefore this writer's opinion that a wise part/document numbering system may not be either totally significant or totally nonsignificant but have "**minimum significance**."

INTERCHANGEABILITY/PART NUMBER/REV CHANGE

Upon noninterchangeable change (after release), the part number must change in one or more characters. And any time a document is changed after release, the revision level must increase—even for a redraw due to a coffee spill on the old vellums. The logic for part number and revision level that must be applied to each drawing or spec affected by a change is in Figure 5.1.

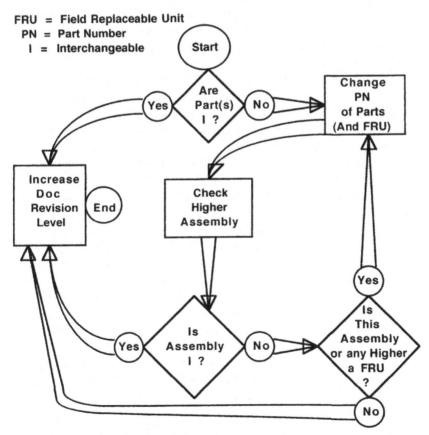

FIGURE 5.1 PN and Rev level change logic.

The logic always leads to one final end—a revision level change. This presumes that the top-level drawing will not change part number—a sound presumption unless the customer demands it.

Note the introduction of the FRU—field replaceable unit. These are **assemblies** which are designated as service items. The reason for introduction of the service item into the logic is to avoid changing part numbers of assemblies which are not spared.

As with most rules, there are exceptions. The parts of inseparable assemblies—welded for example—need not change even if they are "noninterchangeable." See the *EDC Handbook* for other exceptions.

ITEM/DOCUMENT NUMBERING

The document number cannot equal a part number because some documents do not represent parts—schematics, specifications, code, etc. Engineering designs are documented to serve manufacturing to buy or fabricate parts and to assemble parts. Therefore some companies have designated two different systems for numbering parts and documents.

Design of the document number/part number is a critical element in every company. There should *not* be two different numbering systems for documents and parts. The number should be designed to have the document number embedded in the part number.

If your company has already assigned a unique set of numbers to the documents and a different set to the parts, this condition should be critically reviewed. Having a different number for the documents than for the item automatically creates a need for cross-referencing each to the other. Cross-references are often insidious wasters of time.

Doc number to part number and part number to doc number will be cross-referenced several times a day by almost every indirect person in engineering, operations, supply chain, service, and elsewhere. It takes only a few key strokes to do the cross-referencing *but* we are talking about a few key strokes every time any person has one and needs the other—**forever**!

Yes, some CM decisions are "forever" and key strokes do add up! Cross-references are generally evil and this one can be avoided by proper part number/document number design.

A seminar attendee from a medium-sized division manufacturing electronic devices reported to this writer that they studied and calculated the cost of *just* having **a "dash"** (one digit) in the part number resulted in $23,000 in added cost for unnecessary key strokes every year.

Now extrapolate that into several key strokes for cross-referencing PN to doc number and vise versa—everyone, **forever!** To paraphrase ex Senator Dirksen—a key stroke here and a key stroke there—soon it begins to add up to big money!

INTERCHANGEABILITY DISCUSSED

The author finds many feeble attempts to define interchangeability in journals, online, and elsewhere. One PhD wrote as follows:

> *In short, assign a new part number whenever the part has been fundamentally altered; that is, when it no longer fits the previous needs, which are found in both the function and processing.*

Guess it must be up to you to figure out what is "fundamental" or what "fits the previous needs"? Your guess will be as good as any!
Including the "processing"? His elaboration explains:

> *As for the processing, change the part number whenever the design changes affect the processing requirements sufficiently so that the processing times or incurred costs are now different.*

Sufficiently? Cost Different? You decide!

Beware of pundits who try to cover interchangeability in a few paragraphs. In fairness, it is true that the DOD says that when their cost (read price) is to increase that the change must be class I (their poor way of defining noninterchangeable). The FDA is sensitive about process changes—a carry over from their start in drug production. However, neither cost, price, process has anything to do with the real world of interchangeability. Those are arbitrary specification requirements—not a definition of interchangeability.

Most companies say that they follow the form, fit, and function rule. When asked what each term means in the real world, confusion follows. The word "compatible" often enters the conversation. Without defining form, fit, and function precisely, the word compatible is meaningless—not an engineering term— it means you and your significant other get along well.

This analyst emphatically says that; "**We follow the form, fit and function (FFF) rule" is a totally unsatisfactory definition of interchangeability.**

This writer asks the question in his seminars; "Describe a change that *doesn't* affect form, fit or function?" The attendees think on this for a while and someone describes a material only change to a part wherein the material looks the same and has no change of fit and function. Then there is silence! Does that mean that all other changes (except that one) are noninterchangeable? Does that mean that any change in a dimension's tolerance would be "fit" noninterchangeable? Of course not! Thus **the FFF "rule" isn't a good rule at all—by itself**.

A definition and part number change standard needs to be put in place. The factors that need to be present are as follows:

- Fit must be dependent upon the drawing dimensions and tolerances "stack-up"—CAD says "fit interference" or a similar flag.
- Form and function must be dependent upon the product specification or they are open to the whim of the engineer.

- Critical form and reliability characteristics should be in the product spec.
- There might well be an internal spec that is not given to the customer and is used in the determination of interchangeability. Not a practice that this analyst would encourage, however.
- Changes to *meet* the product spec are noninterchangeable and those to *exceed* it, interchangeable.
- After a product is shipped, the part number must change upon noninterchangeable change.

Keeping these criteria in mind, the author's standard interchangeability definition is as follows:

INTERCHANGEABILITY DEFINED

- **Interchangeable**—Two or more items are considered interchangeable if, **in all applications**, they are as follows:

  - Of an acceptable **form** (appearance) to fulfill all aesthetic requirements per the **product specifications**.
  - Of a proper **fit** (physical) to assemble with other mating items per the **drawing dimensions and tolerances**.
  - Of a proper **function** to meet **product specifications** including performance, safety, and reliability requirements.

- **Items meeting these criteria** are completely interchangeable one for the other (both ways) with no special adjustments, modifications, or alterations to the item or related items. (Modify the alterations statement if you wish.)
- **Items that meet some but not all of the above criteria** are not completely interchangeable and are therefore considered **noninterchangeable**.
- **Compatible** is defined to mean that the old is not interchangeable in the new but the new is interchangeable in the old.
- **Whether part numbers or revision levels change** is determined by use of the part number/rev change logic diagram in Figure 5.1.
- **Noninterchangeable changes do not *always* require scrapping or reworking**—disposition of old design parts is a separable decision.
- Only those noninterchangeable changes designated for "retrofit" will be installed into returned goods or field units. "Update everything to the latest revision level" is very expensive and will not be done.
- If the product specifications are *not* clear when processing a change, they should be revised to clarify interchangeability. If the product specifications thus need to be revised, it should be in the same change that raised the issue. If the cognizant engineer chooses not to revise the specification to clarify the issue, that change must be considered interchangeable.

There are, of course, exceptions to the rule. See the *EDC Handbook* for details.

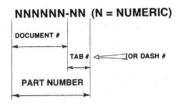

NNNNNN-NN (N = NUMERIC)

TAB SIGNIFICANCE:

- · IF DOCUMENT ONLY 00
- · IF TABULATED DOCUMENT... XX
- · FIRST ITEM (TABULATED
  OR UN-TABULATED) 01

FIGURE 5.2 Ideal part number.

IDEAL PART/DOCUMENT NUMBER

The document number tabulated to make up the part number is the ideal approach. The tabulation has several other benefits including display of similar items on the same document, see Figure 5.2.

Class codes should be excluded from the ideal number unless necessary to "sell" the ideal number. A separate field(s) in the database should be allowed for class coding.

In a start-up company the document number should be embedded in the part number from the beginning. In young companies that haven't done this, the above number convention should be adopted with a carefully managed project. In older companies, it is much more difficult. However, conditions may arise that offer an opportunity to change. For example, acquisitions offer an opportunity to change in any company.

> **Policy: The first practical opportunity should be seized upon for conversion to a single numbering system that embeds the document number in the tabulated part number.**

Caution, it may *not* be wise to change your document/part number design but it should be considered at every opportunity. It should not be undertaken lightly but should be done if practical. Careful testing of the new numbering system before implementation is mandatory. Seminar attendees have reported that poorly planned conversions have brought their company systems to their collective knees.

Numbers that aren't "tabulated," cause multiple documents to be created for similar items. Costs increase due to "reinventing the wheel" is difficult to avoid. Purchasing buys in smaller, more costly lots. One multidivision computer company had 17 different part numbers for the same item because they did not tabulate the part number. What did that cost just in terms of the lost volume-purchase price?

There is also much resistance to changing part numbers upon noninterchangeable change if the part number isn't tabulated. People often memorize part numbers and will resist change of the entire number but do find changing the tab portion acceptable.

Policy: The part number/document number design is a critical foundation element in product manufacturing—conversion should be seriously considered by senior management.

Adding a couple of digits of tabulation to existing document numbers might be in order in any event. A part number without a tab is like a ship without a rudder—for all the stated reasons.

Thus, when the document needs to change to add a new similar item or because of noninterchangeable change, only the suffix (tab) digits need to change. Two digits of tabulation are usually enough.

ASSIGNMENT OF PART NUMBER AND REVISION LEVEL

CM should have a system whereby they assign part numbers to any cognizant design engineer online real time. Minimal data should be required—engineer's name, project, item name, and date. Part numbers can later be reused if needed by doing a used-on search. Part numbers with no used-on can be reassigned by consultation with/notification of the original assignee. See CM Metrics for further discussion.

Revision levels must, as stated earlier, be assigned only by CM as and after the items are released to pilot production. The executive champion must be the enforcer of this policy unless a software system can do the job.

Many companies harbor a fear of running out of part numbers. CM should therefore prepare a graph to track and predict that event as in Figure 5.3.

Action to add a digit or two, or take advantage of the run-out to revamp the part number should be started at least a year before run-out, as it is not easy.

A lighting company learned this the hard way. A significant loss of revenue was experienced when a crash program was implemented.

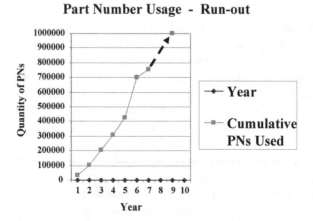

FIGURE 5.3 Run-out of part numbers.

REVISION LEVELS

Some companies mark parts or inventory bins by revision level. The MRP/ERP systems have contributed to this practice by demanding a revision level on every part number—and rightfully so—for rev level is important to those producing parts. Physically separating inventories by revision level and physical marking of parts by revision level are, however, poor and very expensive practices. Such practices are usually the result of poor interchangeability standards or understanding.

> **Fact: Parts don't have revs, documents do!** We often say that the part is at rev J while we should properly say that the part is made from a drawing at rev J.

One should be able to reach into a bin and find interchangeable parts—regardless of what rev level drawing the parts were made from—the blind man in the stock room.

This is not to say that revision levels are not important—if they aren't, the change shouldn't have been done! They are very important to the supplier and the fabrication/machine shop. They are important to receiving inspection. They are important right up to the warehouse.

After that point, the revision level of the document that the item is made from should not be important, tracked, or marked on parts.

REVISION CONTROL

As previously stated—if the CM organization does not have revision control you do not have control—after release to pilot production that is. If any engineer can assign the next revision level you will have chaos.

> **Policy: At and after release to pilot production the revision level must only be assigned by the CM organization.**

This assumes that the pilot units are made by production and that they will be shipped to a customer.

CM is tasked with this responsibility because they must make sure that the proper processes and standards are used. It is all too likely that an engineer will see a need for urgency and bypass the peer review. If you haven't had a quarter million dollars of unusable parts show up on the dock, you will.

This rev level control feature has another very important role. It lets all document users know if their work to implement a release or change can go forward with low risk.

PART MARKING

There seems to be a trend in some companies to mark parts with the part number. It isn't expensive some say—perhaps a quarter! Sure, a quarter to mark the part, then consider what the number one reason for quality control to reject marked parts—can't read one or more digits of the marking. So we now have another quarter to reject the part, another quarter to correct the marking, and another quarter to reinspect....**Quit marking the parts and send the quarters to the writer.** Require that customers and service people use the publications available—which are hopefully up to date with the latest part numbers.

This consultant went through this above litany with a client. When back in the office a letter was received with a big thank you for the significant savings and a quarter taped to the letter. A nice letter was sent in return pointing out that the compliment and the quarter were appreciated but the word was plural—quarters!

If this practice eliminated the need for an illustrated parts list it might be worth-while—but it can't because part numbers change and the latest should be in the IPL.

COGNIZANT ENGINEER

The CM function must know who to accept a release from, who to give a request to, and who to accept a change from. That is unless you are practicing design by committee.

Many others in the company often need to know who is responsible for a particular part/assembly/product design. It is therefore imperative that a list be developed and available that shows the responsibility for design of the product—by product family, type number, assembly number, or even by part number.

> **Policy: The CM manager should work with the chief engineer and others to develop a cognizant engineer list.**

The "list" may be part of the PLM or ERP system or separate. Also add the primary acceptor (cognizant ME, TE, buyer, etc.) for each kind of document or item. The cognizant service engineer, quality engineer might also be added to that list.

This list should be updated and distributed regularly or put online to assure that all who need to know can find out.

APPROVED MANUFACTURERS LIST

The acceptable manufacturers for an item should not be shown on the item document. It is normally a poor practice in this country to let the suppliers know who their competitor is.

Engineering should specify the first acceptable manufacturer for every purchased item. They should do that in the item release document. Engineering, quality, operations and engineering should agree on who the additional acceptable manufacturers are. This process does not need to be controlled by the change process—it can be accomplished by e-mail or a separate system managed by QA for example.

Purchasing will probably wish to specify what *suppliers* are able to furnish the item made by the acceptable manufacturers—therefore the confusing terminology—QVL, qualified vendor list; AVL, approved vendor list; AML, approved manufacturers list, etc. Thus there is potential for two "lists" or confusion as to whether "the" list is for manufacturers or suppliers or both.

The executive champion may have to enter this debate in order to find a company solution.

DEVIATIONS

Most companies have the method of identifying defective parts—usually called a DMR—defective material report. Deviations, waivers, off-specs, etc., are processes for disposing of those parts. For a variety of reasons, they allow some parts to deviate from the design documents.

One outstanding exception to this norm was in practice at Collins Radio Company when this writer worked there. The founder, Arthur Collins, said effectively "we will have no deviations; if the design can change, change it, otherwise return the item to the supplier, rework it or scrap it." This analyst highly favors that approach. The quality of their products was absolutely outstanding as a result of this policy and the founder's attitude. When NASA instituted the "Zero Defects" program, the Collins folks just smiled as they were and had always been in that mode.

Fact: The management's attitude determines the quality of the product.

Seriously consider adoption of the "Collins Policy." Lacking the ability or courage to do that, develop policy that at least prevents the use of the deviation document as a method for making design changes.

Policy: Only deviations will be allowed which are temporary and wherein the design will *not* change.

Thus for a limited number of units or time, the parts will be used and then we will return to the prescribed design. Engineering, operations, and quality should all sign-off on such a deviation.

When defective material is identified *and* the cognizant engineer agrees to change the design so the parts will be acceptable:

Policy: Do not allow use of the defective parts until the change order is approved and released.

This will keep the pressure where it belongs, on engineering to write and process the engineering change document.

If this is not done, the change may never be forthcoming or when the change is written, it may be of a different design than the deviated parts. The configuration of the deviated parts will be found only in the QA deviation files—effectively lost forever. If this policy is not in place, the deviation process will also very likely become a way to make fast changes.

Track the deviations used as a change order and eliminate them. See Figure 5.4.

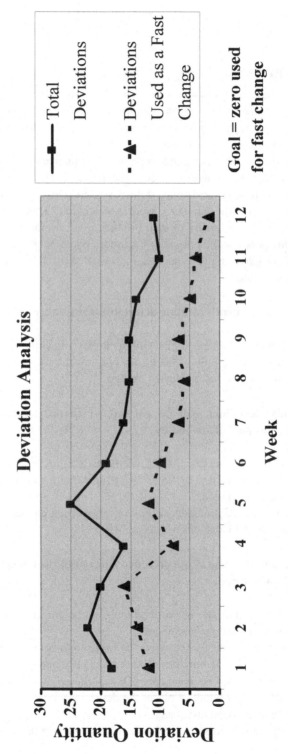

FIGURE 5.4 Elimination of deviations for fast change.

To aid trouble shooters, make sure that the deviation number is put in the revision block of affected documents.

> **Policy: Require that deviations be logged in the affected document revision block. Approved deviations should be sent to CM who would post them to the affected document using the deviation as the authority.**

CM should do this to show a trail to the deviated configuration for future troubleshooting. Your standards should allow this without a change order.

SPARE PARTS

Often a conscious decision as to which items will be designated "service parts" is not made. The result: customers are led to believe that all parts and assemblies are available on a fast turn-around basis. This sets the stage for unhappy customers.

It also promotes an underground stock of items taken from the store room or the assembly floor. Service folks feel that they need to do this to satisfy customer needs. Result: the production schedule is missed because we have robbed Peter to pay Paul.

At a farm implement company the parts issue/production shortages where unusually high. Production delivery schedules were often missed. It was found that the Service people were climbing the stockroom fence at night to claim the needed parts or assemblies. They thought this practice was better than taking parts from the assembly line. Such midnight requisitioning was traced to the lack of a service parts decision/ordering/stocking by the service organization.

In many of this writers consulting experiences spare/service parts are unidentified. The publications list all the parts in the product—an implied promise to customers that all parts and assemblies will be furnished on demand and quickly. Many debates about "priorities" on available parts result.

> **Policy: The cognizant design engineer and service engineer must agree on those items which will be offered for quick sale as spares.**

This decision is, of course, based on the wear, damage, or failure probabilities.

In order to avoid midnight requisitions, missed schedules, and midnight requisitions.

> **Policy: The Service Department must order and stock a minimum quantity of service parts.**

There should be a separate list referenced in the product BOM or the items should be coded "spare" in the BOM. Usually about 20–30% of the total parts in a product will be so designated. Thus robbing Peter to pay Paul can be avoided and 70–80% of the publication cost can be saved. This, of course, presumes that the service organization will stock a minimum inventory of service parts.

PUBLICATIONS

Identification of only the service/spare items in the publication will save considerable publication cost. Thus the time saved in pubs preparation can be used to cure other problems in this often neglected world.

Publications, neither hard copy or online, often do not match the product shipped. R. M. Donovan wrote in *Midrange Enterprise*, "The fact is, most information supplied (to the customer) is excessive, often late and frequently inaccurate."

For many months, a computer peripherals company had a deviation hung on every product shipped because QA observed that the manuals did not match the product. Not only were the publications people not involved in the change process but they were on the other side of the city—in a different division. When this problem was brought to the attention of senior management, a reorganization took place that resolved the problem—no more deviations for pubs.

Some companies are not even aware that the pubs don't match their product. The CM manager or executive champion should make sure that QA is checking the manuals to assure that the manuals match the products shipped.

Often the publications are done in several languages. Huge piles of paper ship with each product. Don't the shipping folks know where the product is being shipped and can't they put the proper language manual with that product? One company did just that and saved a bunch of trees and found customers much more inclined to use them.

All of the above issues with product publications point toward making them available online with QA folks checking them carefully against the product shipped.

We writers often don't use word conservation—the principal reason that reading the publication is often a last resort. Enough said!

NAMEPLATE/SERIAL NUMBER/LABELS

The executive champion and the CM manager should examine nameplates/product labels to assure that all the data displayed are still valid. This analyst examines product nameplates at the beginning of a CM process "walk through/gap analysis." In a surprising number of cases, there are data on the nameplate that no one knows why it is there.

An important bit of data that must appear on the product nameplate is a field that allows one to find out the noninterchangeable change content of the product. This might be the serial number, date code, or "mod identifier" but the "traceability" to the change content must be present on or near the nameplate.

This writer holds that any nameplate, label, or sticker on the product being shipped to the customer must be documented by a drawing and appear in the product BOM. If bar code labels are not so documented they should be removed from the product before shipping.

Do not let the lawyers dictate the need for labels. One popular outboard motor manufacturer—now bankrupt—had about a dozen labels on the product.

Lawyers were running the company. However, when a friend took one into the Canadian wilderness and needed to add gas/oil mixture, do you suppose that one of the labels told him what that mix should be? He also found that the motor wouldn't go into reverse! Could these faults be indicative of why they failed?

TRANSMISSIONS TO CUSTOMERS

Any design drawings or specs given to a customer should be done through CM. They should keep record of all such transactions for the company. If production process documentation is given to a customer CM must insure that a record is available. This is necessary in *all* phases of development and production.

SUMMARY

Put all needed foundation standards in place:

- All general standards need to be in place, debugged, and training complete—PN, interchangeability, etc., before the processes are redesigned.
- Assure that the deviation process is efficient, effective, and is not another way to make design changes.
- See that a method is in place for engineers and service people to define service/spare items.
- Assure that the service folks keep a minimum stock of service parts.
- Assure that publications address in part number detail only the service items.
- Assure that product nameplates are documented, contain only information required, and contain a field which will allow tracking of changes when necessary.

Chapter Six

Signatures on CM Documents

Signatures represent responsibility. Signatures on design documents are extremely important because the company's product is embodied in the design drawings and specifications.

Use of the word "signature(s)" in this book applies to physical signing of a document or a secure online "approval."

Too few signatures may mean inadequate review of the document. A multitude of signatures are widely attributed to long process time. Many say they are the single biggest contributor to excessive process time. As a famous comedian said "Everybody wants to get into the act."

When this author did product design work, he signed the new document, a checker signed, the boss signed, and his boss signed. The author was very comfortable with this process because he knew that if something went wrong "they would not fire all of us!" No one from operations (the primary customer) signed. This orthodoxy is all too typical in product manufacturing.

INVOLVEMENT VERSUS SIGNATURE

The temptation is strong to allow any function who *might* be affected by *any* change to sign every change—but that is a touch of insanity!

The question of "who signs" will be addressed in each process. For now, realize that many people/functions should be **involved** in the process but very few need to **sign.**

> **Policy: Approvals are much like "Ham and Eggs," where the chicken is involved but the pig was committed.**

A personal-vehicle manufacturer liked this analogy so much that they began to label themselves as "pigs" or "chickens" in the CM processes.

Each member of the change team should have their responsibility carefully defined in a CM Standard—manufacture-ability by the manufacturing/industrial engineer, inspect-ability by the quality engineer, test-ability by the test engineer, service-ability by the field engineer, etc.

49

Configuration Management for Senior Managers. http://dx.doi.org/10.1016/B978-0-12-802382-2.00006-4

One electronic device manufacturer had 11 signatures on each release, change, etc. A seminar attendee reported that they had 15 signatures on changes. Do all of those people and engineers need to sign releases or changes—as you will see later—No—just the pigs.

Signing the release or change document is typical in industry. Are the responsibilities for the manufacturability, etc., on the drawings and specifications or on the release and change documents? The answer seems obvious to this analyst.

> **Policy: Required signatures shall be on the document(s) to be released, not on the Release Form. The technical signatures required shall be on the change markups—not on the change form.**

On changes, production control should be directly involved in the process to set effective dates/serial number. You may wish them to "sign" the change form for that contribution—or not.

Since CM is (should be) in control of the processes, they need not sign because they can stop the processes at will. You may wish to have them sign the release and change documents or merely enter the responsible CM tech's name.

The use of face-to-face meetings, well chaired, is preferred by this analyst whether you have an online or manual process. If you have fast, accurate processes with online process and signatures, then face to face may not be needed.

If you have CM "in the middle" attempting to gather technical signatures, you will have a fast runner with quick fingers or a slow system.

ONE AUTHOR AND ONE ACCEPTOR

Most companies have several engineering department signatures on release orders and change orders. Curiously, design drawings and specs are seldom signed by the internal customer and no one typically signs a parts list. Go figure!

For new documents, the best practice would have one author—the responsible design engineer and one acceptor—a technical representative from the primary internal customer. The object of this approach is to make responsibilities very crisp and clear—one responsible for the design and one for the manufacture. Each should call on others in or related to their discipline as needed before they approve.

For most design documents, the primary customer is manufacturing. For a product specification, it is probably marketing. For a test document, it might be quality assurance or test engineering. The operations management might assign a single function (such as ME or IE) to represent all departments—including even the supply chain. A manufacturing engineer or industrial engineer can be cognizant of both make or buy issues.

Based upon the company culture or organization structure, a separate signature representing the supply chain may be necessary when a purchased item is involved.

This practice need not take away the design managers' right to manage. They can still review designs before release as needed. It merely places the burden for best practice where it belongs—with the cognizant engineer and the primary internal customer representative.

The same practice can be applied to the change process. Some changes would affect multiple documents and thus require a different signature as acceptor on the applicable document.

Thus, there is a need to document the specific author and acceptor responsibilities. This takes us back to the cognizant engineer list. That list, as previously stated, would contain the design, manufacturing, and others as needed for signing documents according to your policy.

> **Policy: Limit signatures on documents to an absolute minimum, but at least one author and one acceptor should sign.**

It is universally believed by this consultant's clients and seminar attendees that more signatures do *not* improve the quality of the outcome.

Consider the paradox wherein several people sign a drawing, drawing release, or change, while no one signs the parts list.

SIGNING THE BOM OR PARTS LIST

This analyst has never witnessed a company that signed parts lists or BOM. It is generally implied that someone is responsible and assures the accuracy—not always a good assumption. At the very least, the standards should clearly state that the CM organization is responsible for the accuracy of the design data in the BOM.

It can easily be said that the cognizant engineer should be responsible. Certainly, the engineer is responsible for the initial release information and the delineation of changes; however, the accuracy of the input (and check of the output) should be CM's responsibility.

An online approval of a "proof copy" might be a way of assuring accuracy. CM would use a different person to check the output than made the entry.

Zero signatures on the parts list/BOM is acceptable if the responsibilities for the generation, input, and checking of the data entry are carefully specified in a standard and controlled by CM.

JUST IN TIME

"Just-In-Time" manufacturers usually have a bell/buzzer/light that *anyone* in the production operation can use to shut down the line. Any assembler can shut down the entire assembly line. It gets attention and solution to every problem very quickly. This philosophy can also be used in the CM processes.

> **Policy: The CM processes should always contain a provision for anyone *directly affected* by a document, to stop the process by notifying the CM manager that something is amiss.**

There should be a *time limit* placed on this "stop order" resolution. It should be incumbent on the CM manager to analyze the issue and determine a resolution promptly.

This provision allows empowerment of those affected and thus should replace the need for many signatures.

WHO OBTAINS SIGNATURES?

Another very important issue: who obtains the approval signatures? Often, this task is delegated to the CM people—very often by the design of the OLM online process. There are two kinds of "approvals"—technical and administrative. Technical issues require technical answers and discussion—the cognizant engineer should have the knowledge and responsibility to deal with other technical folks.

> **Policy: Technical approvers on drawings, specs, code for release, and redline changes should be obtained by the cognizant engineer.**

This approach avoids the middle man, back and forth that will otherwise result.

The administrative signatures should be "obtained" by the CM function. This will also allow complete separation of the technical from the administrative in all the processes.

Online systems often do not specify who should be expediting the required approval(s). The standard associated should be very clear about this issue if we are to have a sense of urgency and fast, accurate processes.

QUALITY ASSURANCE ROLE

Quality assurance folks often sign documents, releases, and changes. This makes them part of any problem that occurs. How can they critique a process in which they are intimately involved? This is somewhat similar to 100% inspection of all incoming parts. Is it not better for the quality representative to stand back and sample/observe the processes and report on process issues?

The same reasoning would apply to CM. If they control the process, there is no need for their approval. At key points, after they assure that the proper process has been followed and the other approvals are present, they would move the process to the next step. With such control, there is no need for CM to "sign," but the responsible CM technician should be identified.

TECHNICAL RELEASE

The point in the process for obtaining required technical signatures—releases or changes—is critical. This is an event in the release and change process by which technical issues must have been resolved. Only administrative issues remain. Those technical issues include the following:

- Manufacturability-/Produce-ability review;
- Inspect-ability review;
- Test-ability review;
- Design complete as evidenced by all documents or markup available and meeting standards;
- Customer approval or review if required (or a conscious decision to proceed without same and clear responsibility for customer follow-up);
- Cost of the item to be released or the change cost if necessary.

The administrative issues are addressed after technical release—those include the following:

- Updating of the PDM/PLM system;
- Updating of the ERP system;
- Sending documents to customers if required;
- Finalizing assembly and fabrication processes;
- Placing/modifying shop orders;
- Placing/modifying purchase orders;
- Determining when to make the change—date/serial number effective.

The separation of technical issues from administrative issues allows us to divide the release and change processes with clear responsibility assigned.

- Before technical release, the cognizant engineer is responsible.
- After technical release, Configuration Management is responsible.
- After technical release, the operations folks can start their tasks for implementation with very low risk.

If something is wrong with the item released or with the change, another change order will be required. The engineer must write it if it is a technical issue, and CM must write it if an administrative issue.

Policy: There must be a complete separation of technical issues from administrative issues in the CM processes.

This practice fosters "doing it right the first time." This is critical to the speed and accuracy of release and change processes.

We will discuss a similar principle for the request process in the corresponding chapter.

SIGNATURE ALTERNATIVES

The primary alternatives for signing CM documents are the following:

1. The one author–one acceptor approach.
2. Each VP could be allowed one signee—engineering, operations, supply chain, and service if affected.
3. One signature for each affected function—ME, TE, QE, Pubs, Service, Purchasing, etc.

Your choice should always be accompanied by the just-in-time stop order. An ideal signature process for your company might use combinations of these alternatives. Specific method will be suggested for each process.

SUMMARY

We want many functions involved in the CM processes, the more eyeballs the better, but

- Realize that signatures add time to the processes without necessarily adding value.
- Minimize the signatures involved in each process.
- Devise a method wherein anyone directly affected by the process/ documentation has the ability to stop an action without a signing privilege.
- Make sure people understand what their responsibilities are and what their signature/approval is for.

Chapter Seven

Process Improvement

If we are to stay ahead of the competition, processes must be continually improved. Details for process improvement are covered in the *EDC Handbook* in Chapter 12. Either continuous improvement or reengineering or both techniques can be used. The following is a summary for reengineering a process.

1. Put key metrics in place in order to verify improvement.
2. Write standards on elements of the system which you know to be basic elements or subjects—not likely to change with process redesign. We discussed the most important of those subjects in Chapter 5.
3. Get each standard approved and implement those standards one at a time.
4. Make sure that the standards are short (1, 2, or 3 pages) and cover only one subject. Longer standards are probably covering more than one subject and should be subdivided.
5. Flow diagram the current processes you wish to reengineer—release, BOM, request for action/change, or change order.
6. Flow diagram a new streamlined process.
7. Write standards for that process and "work" them to approval.
8. Implement the new flow—probably on one product. Be prepared to revise the processes and standards as needed.
9. Observe the key metrics to assure that improvement occurs. Don't expect instant results—it may take a few weeks to see improvement.
10. Implement the new work flow on all products.
11. Repeat the above for each process.
12. When all processes have been streamlined, write a system summary standard that defines/summarizes how the system works. This standard will be the one that you give to new employees, customers, agencies, suppliers, or anyone else who asks, "How does your CM system work?"
13. When all processes are working efficiently and effectively, get ISO certification—or not.
14. Then proceed to continuously improve all the processes.

These steps need not be done sequentially—some can be done in parallel.

When doing continuous improvement, do steps 1 through 5 and then make the new process work flow diagram, but implement it in manageable pieces.

Configuration Management for Senior Managers. http://dx.doi.org/10.1016/B978-0-12-802382-2.00007-6

The goal for process improvement should be for speed—without "hurrying up to do it wrong" and thus over and over.

> **Policy: The first step in process improvement is to measure the quality, speed, and volume of the current processes—release, BOM, request, and change.**

Thus the major goals of a process improvement program should be speed with improved quality of releases, BOMs, requests, and changes while taking into account the volume of those actions—including work in process.

Specific metrics/measurements will be presented as part of each process discussion but let's examine the most frequent problem area—process speed.

PROCESS SPEED—CASE HISTORY

A money-cow division of a Fortune 500 company achieved fast, accurate release, request, and change processes with:

- a dedicated executive champion,
- a three-person team,
- a cooperative CM manager,
- an old (but good) MRP system,
- CAD, and
- *Completely manual* CM processes.

Many aspects and results of that 2-year project are discussed more thoroughly in the *EDC Handbook*. The summary of the results is as follows:

- They were able to release a new item/document in about **three work days** average.
- They processed the average request to change in about **four work days**.
- They were able to process a change to the design documents in **a little over and average of five-work-days.**

The time was measured from technical release to update of all documents and systems complete. This was taking place while the volume of actions was increasing.

Note that in many companies there is not a separate request process—thus comparisons may be difficult to make. Also, not all changes were preceded by a request.

Specifics may help in making comparisons for those operations that measure the speed of their processes:

- The **release** process was measured from "design complete" (docs technically signed and given to CM), until the systems are updated and the new design documents were available to view/copy/print.

- The **request** process was measured from the receipt of the request in CM until the requester was notified as to the disposition of the request—rejected or accepted for the next iteration of the product.
- The **change** process was measured from the point of technical release (markups signed and the change form technically complete) to the point of master documents being updated and the systems updated. This includes the initial plan for the change effective date.

The design and development time and the implementation time are not included in the stated averages but were also measured. In their case, those average lapsed times also went down as the processes were redesigned.

Yes, this was done without purchasing any more software or modifying the legacy MRP.

CASE STUDY CHANGE PROCESS REENGINEERING

This company recognized that the change process needed to be addressed first because engineers will resist releasing any document when the change process is slow and painful. Besides, the change process was, as in most companies, the most wanting of attention.

That company had measured about 40 work days to redesign and develop the change, about 40 work days through CM, and about 40 work days to implement the change. This analyst called it the "biblical" change process—40 days and 40 nights. The middle 40 was driven down to about 5 (when no request was involved), the "design 40" and the "implementation 40" both went down a few days. Effectively the **total change process was reduced about 40 work days.** Thus,

- The customer saw the fix, feature, or option 40 work days sooner.
- 40 work days less scrap or rework.
- Units that needed to be retrofit reduced by 40 work days of production.
- Recalls affect 40 work days fewer units.
- When there was a real per unit cost reduction—$X per unit—multiply that figure by 40 work days of production to calculate the real cost savings.

That total turnaround time reduction from identification of a design/documentation problem until implementation in manufacturing was accomplished while the quality of the design and change documents improved. It was not a result of hurrying up to do things wrong. In fact, the reduction of "correction changes" was a significant contributor to cutting the process time.

It must also be kept in mind that the request process was separated from the change process and it took an average of four work days—thus *when a request was involved*, the new total average process time was a little over nine work days.

Again, this was achieved without purchasing a single dollar's worth of new software or modifying existing software and with a totally manual change process—paper, telephone, face-to-face meetings, and walkabout expediting.

PROCESS TIME EXPECTATIONS

Measure the throughput time and compare that with what this analyst sees as reasonable expectations. After your improvement project is complete, you should expect, with existing software, the average process time to reach the following levels:

- Release time to be two to four work days
- Request time to be three to five work days
- Change time to be five to seven work days

There will be much more harmony among the people involved, process outcome quality improvement, and other benefits as outlined in "The Dirty Half Dozen."

With these thresholds reached, addition of or changes to process software apps might be in order to further improve the efficiency and speed of the processes.

Specific time and quality metrics will be explored as part of each process discussion.

TEAMS

Teams are a necessity in all the CM processes simply because so many functions are or can be affected.

If your CM process using the PLM system is working to these expectations via an online method, you may have no need for team meetings. If not, face-to-face team meetings should be given careful consideration in the new process designs—at least until a culture with a "sense of urgency" is developed.

Tech review meetings in the release process are common practice. Sometimes they are not held often enough but new documents are usually reviewed on a team basis—although not always in item lead time.

Some release processes may function online within our expectations because team meetings are already in place and function well. If they aren't within our parameters, frequent team meetings in the release process should be implemented immediately.

If the current release process is fast because there is no "acceptor" signature required, a different issue enters the picture as previously discussed.

There should be a high level management team to review design change *requests*. This should occur two to three times a week. The best process this writer has witnessed was a team of the vice presidents of engineering, operations and the supply chain chaired by the CM manager. They were able to sort out the wheat from the chaff very efficiently.

Many folks who have been involved with "board meetings" in the change process will cringe at the thought of returning to live meetings. This is because they remember those meetings as:

1. **Poorly chaired**
2. **Taking way too much time for many people**
3. **Too infrequent**

4. Too late in the process
5. Too many people with signing authority
6. Having ill-defined team member duties
7. Trying to solve problems in the meeting

Those CM change "board" meetings were normally held once a week and often go on for 2 or 3 h. They were poorly chaired. Folks tried to solve problems in the meetings rather than outside the meeting room.

If meetings are held only once a week, *any issue could cause a weeks delay.* Thus the throughput time will be measured in weeks rather than days.

Policy: Short, frequent, well-chaired team meetings are mandatory for quick change action.

The teams involved must meet at least three times a week if not daily. If done daily, the delay (when problems occur) will be 24 h, not a week. Half-hour meetings are normally adequate—after the backlog has been worked off.

The change team should be able to reject requests back to the request management team for reconsideration.

The meetings must be well chaired. See the *EDC Handbook* for details. One company used a high table without chairs in their CM team meeting room. It was not totally surprising that the meetings were short. Holding the meeting just before lunch will also assure brevity.

CHANGE BOARDS

The typical Change Control Board (CCB) meeting is held after the engineer thinks they are ready to release the change. This is too late in the process for other folks to first see a proposed change.

Bad things happen if the first time the technical reviewers see the change is at a point too late in the process. This consultant has witnessed high emotion, swearing, and even physical confrontations in "change board" meetings. A simple issue like a tolerance, which the supplier cannot meet, becomes a major issue. This happens because the design engineer simply thought that he or she was done, went on to their next problem, only to be embarrassed or frustrated by some issue at the next weekly change board.

Technical review team must begin very early in the change process. The initial review should be right after the request is approved or the change project is begun.

Policy: The design engineer should prompt team discussion on any design issue at the first team meeting after becoming aware of the problem or after the request has been approved.

The request/problem/issue should be discussed at further meetings as required. Allowing the team to inject their 2 cents worth early in the process will pay multiple dollar and delay dividends.

Use redlined documents to define the change because it is the best way to see both the current and the proposed conditions. When the engineer has redlined drawings and specs available, they should be viewed in the next team meeting. All technical issues must be satisfied prior to considering the change ready for technical release to CM.

Similar team activities should be working in the software development and change efforts. Consideration should be given to having the product CM organization track the software change requests.

The software teams would probably be different people/functions than the "hardware" teams but the tracking of requests and the release activities are very similar.

FORMS

Much to do is sometimes made of the number of forms in the CM processes. This analyst sees no relationship between the number of forms and the efficiency or effectiveness of the processes. The processes must make sense, be documented, fast, accurate, efficient, effective, measured, and well understood whether 3 or 13 forms are used.

Whether online or hard copy, forms must have clear form instructions. See the *EDC Handbook* for form and instruction details.

If the same form is used for release as well as change, then two form instructions should be written. In this case, two separate forms are probably wise.

Pop-up instructions for an online form are an excellent practice, providing CM writes the pop-up instruction.

CUSTOMER ORDERS

The ultimate goal of CM processes should be to satisfy the customer's needs. The customer order process should be considered for inclusion into the CM arena. It is often troublesome for reasons mentioned earlier.

The sales people often use different product identification than engineering and manufacturing. Result: someone in order entry tries to convert the customer needs into engineering and manufacturing numbers. Thus errors are made that cause the customer to receive something different than they ordered. What is one unhappy customer worth? Remedies were discussed in the part number discussions.

Primary among the critical customer requirements is to receive their order in the promised delivery time. Thus one of the first measurements that should be established in many companies should be on-time delivery.

One metric that would apply to a make-to-order company is shown in Figure 7.1.

Certainly late delivery is damaging. Sometimes early delivery is also frowned upon. The time to process an order from sales to order entry and order entry to operations should also be measured—example metric in Figure 7.2.

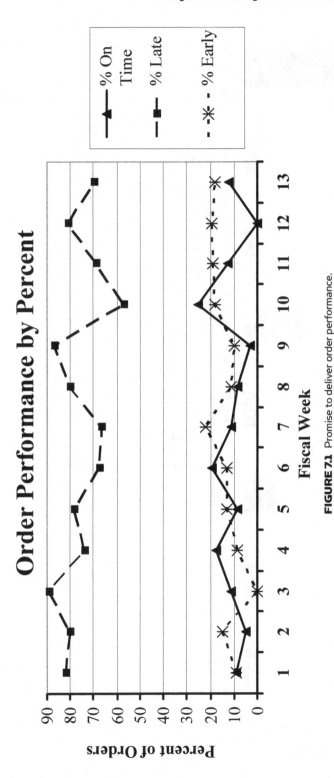

FIGURE 7.1 Promise to deliver order performance.

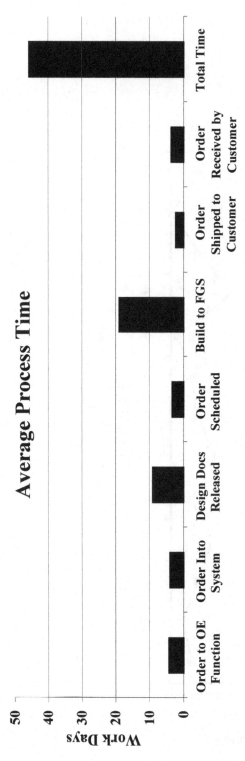

FIGURE 7.2 Order processing time.

This metric will allow evaluation of the process in terms of time in each phase as well as total time.

Such metrics should be established early on in the CM tool kit and watched carefully as process improvements are made to assure that the promise to deliver time is not increased (do no damage) and shortened where possible.

A mechanical device company adopted CM process changes that included modular BOMs and cut their promise to deliver time from 26 to 6 weeks. They had to put common assemblies into work in process to achieve that reduction but they beat up their competition for some time to come.

An injection molding company improved their on-time delivery from the 50% level to over 90% in three plants by implementing a new order entry software, a new CM function, and an ERP system—without additional manpower. The project took 2 years to start-up but within the next year the on-time delivery goals were met.

BACKLOGS/WORK IN PROCESS

It is mandatory that the backlogs (WIP) of orders, releases, requests, changes, etc., are driven down to help attain the desired throughput time.

If your operation processes 30 change requests per week on the average, and your goal is to achieve a five-work day (1-week) request process, then the backlog must be 30 requests average, at any point in time.

The same crude math method can be applied to any process—deviations, orders, releases, requests, changes, etc. The chart and graph in Figure 7.3 illustrates how the throughput time can be predicted by this method.

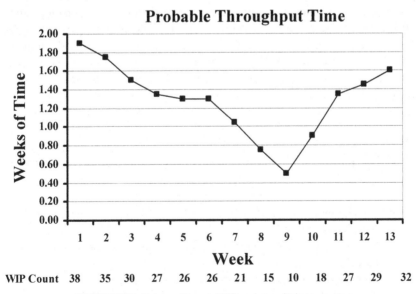

FIGURE 7.3 Process time via WIP and completions count.

You will notice that the process speed decreased over time to a half week, but then increased back toward the early poorer performance. If one half-week throughput time was once achievable, it should probably be sustainable—analysis is required to find causes and solutions.

In order to be precise, each document would need to be time measured from start to complete. This would get the word "probable" out of the metric.

Of course the amount of time a form lays in backlog (aging) has a bearing on this throughput time math, but that can and should be watched with aging reports.

Occasionally, the reduction of backlogs is all that is needed to achieve a fast process. This analyst found in one mechanical sprinkling device company that the change process was mostly sound, but that changes were allowed to accumulate in the process beyond reason. The chief engineer inaugurated a backlog reduction program that was unique in this analyst experience.

Every engineer in design engineering as well as operations (VP of operations agreed) was given a few changes and had to come in on Saturday mornings until the backlog reached the target.

Of course the process had to be carefully mapped for folks to follow. Sign-off or rejection took place in real time in the offices/hallways/restrooms. Very few engineers had to come in a second Saturday but none had to come in a third one. Drastic solution but it was an instant way of instilling a *sense of urgency* message that was well understood and the fast process time was maintained.

AGE REPORTS

Lists of releases, requests, or changes are usually kept by CM. Those logs or lists should be displayed in age sequence—oldest on top. The person responsible for the next action should be prominently displayed. Copies should be sent to the team and management. This simple technique can work wonders simply because no one will want to be on the top of such lists. This is the method the above company used to maintain their change throughput time at the desired level.

SUMMARY

Choose the process to be improved and prepare a plan for your process improvement activity:

- Having an executive champion involved with process reengineering is most likely to bring success.
- Without an executive champion, continuous improvement is probably the best approach.
- If an improvement team is involved, keep it small.
- Assure that metrics are in place.

- Set reasonable goals for process time and accuracy.
- Seriously consider returning to face-to-face meetings in this process at least until a sense of urgency is deeply ingrained.
- Study the WIP and volume completed to predict the throughput time.
- Sequence detailed work in process reports by age.
- Reduce backlogs of work in process.
- When this process is working to expectation, go to the next CM process.
- When you have all processes working to expectations, set new goals and do it again.

Chapter Eight

Release Process

In order to promote an orderly and rapid product development, a well-conceived process needs to be put in place. The speed and effectiveness of the product release process is critical to capturing the market "window." Engineering's "product" is specs (including software code) and drawings. They allow repeatable production of the company's product. Thus the need to control and streamline the product documentation release is **strategic**.

Release of an item from engineering to production and the market is generally done in phases—the object being to have a systematic, orderly, measured, and effective process that minimizes risk while acting with speed.

ENGINEERING AND COMPANY PHASES

Many chief engineers have a normal product development process which they have implemented to achieve the best design results—but what about the rest of the company? They may not be interested in many of engineering's detail steps and the documentation is often an afterthought in those processes.

There is sometimes no correlation between the engineering defined phases of product development and the "company phases." Result: the nonengineering people are not in sync with the engineers and their management. The process is slowed and the market window gets smaller. Bad parts result. A significant portion of the "bone piles" in manufacturing are traceable to release phase confusion. Hours of clarification discussions result. Finger pointing happens frequently.

Most engineers are focused on the product. If they have a working model or tested breadboard or functioning program, they see a significant event. But the rest of the company deals with the design documentation.

There needs to be "company phases" as opposed to "design and development phases." Every company needs to determine the major "phases" of product release—a natural task for the executive champion.

Definition: Release is a handoff of the product design documentation from one major phase in the product lifecycle to the next.

Notice the word "documentation." It is not the purpose of configuration management to delve into the product design (CAD) or software design (SCM)

Configuration Management for Senior Managers. http://dx.doi.org/10.1016/B978-0-12-802382-2.00008-8

processes, except where it is important to the interfaces with production, service, customers, etc.—that is, their output of design documentation.

One medical device company was having fits trying to agree on what the significant phases were in their environment. Many hours had been spent debating this issue before this writer got involved. After an hour in an early meeting with a dozen middle management folks, it was obvious to this writer that they simply were talking different languages. The people involved were from different backgrounds. They used the same word to mean different things. After defining terms (breadboard, prototype, pilot, release, approval, etc.) very carefully we had a breakthrough into daylight. Talking the same language is very important. Get a dictionary of company terms developed early.

The release process must cover release of a design document, an item and its documents, an assembly and its documents, code and its documents, combinations of these, and the end-item product and its documents. It must cover all major phases of the product lifecycle.

Release of a new item (a new part number) *for a change* should be done as part of that change order and will be treated in this work as part of that change—not a separate release action—in the chapter on Change.

So let's first define the major phases of release.

PHASES OF RELEASE

The major phases usually are as follows:

1. Contracting/planning/definition
2. Design and development
3. Pilot/sample/preproduction
4. Production
5. Obsolesce

These major phases vary from company to company. The nomenclature is especially variable.

Sometimes a contract or project may be only for the design and development of pilot/sample units for delivery to a customer (design to order)—resulting in phases 1 through 3 being applicable. Sometimes other circumstances may add or delete a major phase.

Policy: The executive champion should assure that the company's normal major phases are logical and specified in a CM standard and named according to general usage.

Exceptions to the phase plan for any given product development should be approved by senior management.

The phase that a given item development is "in" (valid for) should be readily visible to all concerned.

Phase	Documents	Systems
Definition	Rev dash	X
Design and develop	Rev date	E
Pilot/Sample	Rev numeric	S
Production	Rev alpha	P
Obsolescence	Rev OBS	O

Policy: The "company" phases should be readily visible on both the documents and in the systems—and thus reflected in the parts lists and bills of material as well as on drawings and specs.

Thus, a rule of thumb for coding the documents and the PLM and/or ERP system item master file needs to be established. The revision level is normally used for the documents. A single letter code for the system(s) is usually adequate.

For example:

We can see this phase relationship for the middle three phases in Figure 8.1.

The rule of thumb used isn't particularly important but it should be consistent over the years. The goal is to make it perfectly obvious to all involved as to what phase a document (or an item made from the document) is valid for—whether they are viewing the systems or the documents.

Thus a buyer won't buy an item in quantity for production from a document/system that is valid only for pilot. The fabrication folks won't make pilot quantities when the document/system is only valid for design/development, etc.

CM must assure that the phases identified on the document and in the systems are "in sync."

RELEASE IN LEAD TIME

Rarely can the product design documents be "held" in engineering for the ideal design/development effort to be complete and then released en mass. Reality of customer and company schedules demand that the release happen piecemeal—in lead time to buy and build—if we are to meet the planned pilot or production schedule.

Often engineering doesn't want to release those parts because the change process is so painful and/or slow. Sometimes the hesitation is simply because they don't understand the need.

Operations/production control often create a "planning BOM" for long-lead part ordering via ERP without engineering release. The result is that operations is taking the risk for our late/slow release process. The planning BOM offers huge potential for buying/building an item that has known problems in the design or documentation—the result—more wasted parts and labor costs. The risk is solely in manufacturing while it should be a shared (company) risk!

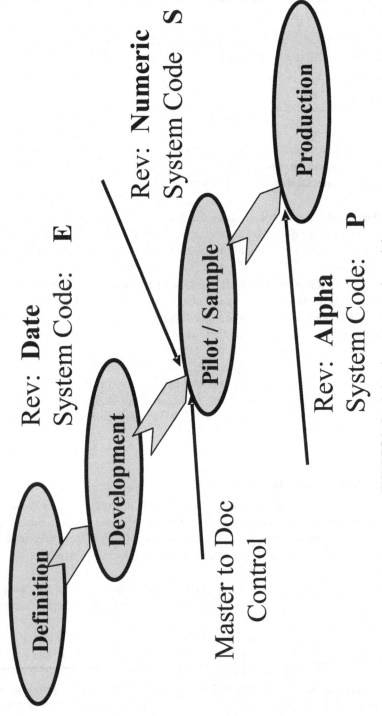

FIGURE 8.1 Correlation of revs, system codes, and phases.

The process must therefore be tuned to releasing an item in lead time per the build schedule.

Policy: Operations (usually production control) must furnish lead time estimates to engineering for all new items in a new product development. Engineering must release those items to buy and build in lead time.

Failure to do this will result in missed schedules. In fact, it is rather easy to predict a day-for-day schedule slip (or significant cost hit to recover) from a delayed release of any item.

Policy: The production control manager must notify executive management of each item which has not been or will apparently not be released in lead time.

Some extraordinary effort and cost can sometimes make up for an item missing the lead time release. However, that makeup effort can't happen on too many items or the schedule will likely slip. The executive management needs to be aware of each such case.

The executive champion should give production control and CM guidelines for how and when to make senior management aware of slips or pending slips of the schedule.

The engineers may release longer lead items out of sequence if they wish or should if they are associated with a short lead item.

CM should enter/update the codes in the ERP and PLM as appropriate.

LEAD TIME RELEASE PERFORMANCE

Production control should track the release performance. Lacking their cooperation, CM should prepare a chart of the releases—in the case of Figure 8.2 it is done in long lead time sequence and is obviously closing the barn door after the horses got out. A good method of highlighting the problem for the next product release however.

In this example, as you can see, there were eight items released after the needed release date. Four were released ahead of the required date. Releasing ahead of schedule is okay, providing we didn't rob Peter to pay Paul. That is, perhaps we put resources into item(s) released ahead of the needed date which could have been put to those items released behind the needed date.

TEAM IN RELEASE

The lack of or poor utilization of a cross-functional team results in designs that are costly or even impossible to manufacture. Fabrication, assembly, test, or packaging time may be increased and/or disrupted because their needs have not been considered "up front" in the process.

Item	Item Description	Lead Time in Weeks	Work Days After Need	Work Days Before Need
1	Base Casting	42	10	
2	Left Forging	38		9
3	Rear Casting	36	9	
4	Molded Front Panel	35		12
5	Right Forging	33	11	
6	Molded Left Brace	30	5	
7	Molded Right Brace	27	1	
8	Molded Side Panel	23		6
9	Molded Back Panel	21	7	
10	G2 Gear Stamping	18	4	
11	G1 Gear Stamping	15	2	
12	Machined Gear Shaft	13		5

FIGURE 8.2 Lead time release performance.

Most companies designate a team of folks to make the new product development process fast, effective, and accurate. The intent is for the team members to:

- Assure the manufacturability/produce-ability of the design.
- Do their many tasks, as much as possible, in parallel with the design effort such as fabrication and assembly processes, tool design, etc.
- Bridge the gap that may exist between engineering and the rest of the company.

The teams are typically made up of technical folks from engineering, operations, and service. Certainly production control should be represented for lead time coordination. It is also likely that someone from the supply chain should also be involved although a good ME/IE might represent the buyers and suppliers.

The chair of a given team might be the project engineer or, since much of the work is administrative, the CM manager or their representative might well be the chairperson. CM should certainly be on the team to assure standards are followed.

The executive champion should look in on the team occasionally to make sure it is functioning smoothly, that policy is understood and to help remove obstacles.

Considerably more discussion about teams can be found in the EDC Handbook.

SIGNATURES ON RELEASES

As already discussed, each *design document* should be signed by the cognizant engineer (author) and by the primary internal customer (acceptor).

Policy: The release form itself doesn't need to be signed by anyone except the cognizant engineer providing the proper signatures are on the documents to be released.

CM will assign the proper revision level and system coding. CM will not process the release unless the proper process has been followed (team review, lead time, coding, etc.). The CM representative should be identified on the release form and you may choose to have them signed.

The team should not and need not sign but should each have the authority to stop the release by written notification to CM as to the issue involved—process by exception—in the JIT fashion.

RELEASE PHASE CHART

A number of policy issues can and should be specified in a phase-chart standard. Let's take a generic company that has three phases (ignoring planning and obsolescence for the moment) and decide to call them development, pilot, and production. The chart in Figure 8.3 will address numerous company issues by phase:

Notice that there are no "ifs," "ands," or "buts" in the chart. Each line item or "issue" needs to be carefully analyzed in terms of minimizing control, while

Item / Issue	Develop	Pilot	Production
Name of Units	Prototype	Pre-Pro	Production
# of Units to be Built	3 - 6	20 - 30	Per Schedule
Build by	Engineer	Pilot Mfg	Mfg
Serialized	No	under 100	over 100
Testing	Eng Lab	QA	Prod test
Ship to Customers	No	If upgraded	Yes
Master Doc In	Eng	CM	CM
Signatures on Doc	None	Eng	Eng & ME
Revision Level	Dash/Date	Numeric	Alpha
Systems Status Code	E	S	P
ECO to Release	NA	Yes	Yes
Change Control	Engineer	Informal	Formal
ECO Signatures	NA	Eng/ME	+ Field & PC
Updates Master Docs	Eng	CM	CM
Sign Updated Master	NA	CM	CM
I/PN chg rules	No	Yes	Yes

FIGURE 8.3 Phase release chart.

also minimizing risks. This chart and the rules for release should be incorporated into a standard.

> **Policy: The executive champion should assure that CM adds to/tailors this policy chart as necessary and develops a standard with accompanying rules for your enterprise.**

Normal expectations for every item, assembly, code, or product need to be clearly understood by all involved. Senior management can take exceptions as needed. **This is a critical standard for the enterprise.**

RULES FOR RELEASE

The phase chart should be accompanied by some rules:

- Any new item to be released must be accompanied by its associated documents if they are new docs. The CM function should assure that this occurs. For example, if we are to release a new part made of an existing material, only the new item drawing need be released (referencing the existing material spec). If a new item made of a new material is to be released, then the drawing and the material spec must be released together or the material spec first.
- If a new item requires a test in operations to assure its worthiness, then the item drawing/spec should be accompanied by the test spec.

Sometimes test specs are needed for purchased items, either for the supplier or receiving inspection—they should accompany the item release.

- An assembly pictorial drawing should be accompanied by its associated parts list, and all the items in the assembly must either have been released or must accompany the assembly pictorial release.
- The top-level product likewise must have all parts and assemblies, test specs, etc. released before or with the product release.
- If UL or other certification is required, that must be obtained before the end item is production released.
- The product spec must be released one phase ahead of all other documents.

These rules need to be followed for pilot release and for production release. Thus the BOM will/must be released from the "bottom up."

If the product spec is not the first document to be released: What are we designing? What do we test for? What will we produce?

PRODUCT SPECIFICATION RELEASE

The product spec may take a variety of formats and is sometimes not a controlled document—but should be—*as it is the company's most important design document.*

Too often the product spec is bouncing back and forth between engineering and marketing with red marks all over it—a bad but not an unusual situation. Some level of agreement is necessary in order to guide the team in subsequent development.

This consultant has seen a product spec unsigned, unreleased, dripping with red ink when the product is entering production—a very risky situation indeed!

The product *reliability* expectations should be reflected in the product spec even if you do not intend to publish those expectations to customers. This is necessary to make the functional aspects of the spec understood.

Likewise the product *safety* expectations should be in the product spec. This is necessary for functional completeness and liability protection. The number of companies which violate this principal is staggering—see Figure 8.4.

The product specification should be given a document/part number in the planning phase and subsequently released one phase ahead of all other product design documentation.

Product Safety Specifications

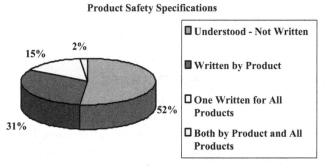

FIGURE 8.4 Product safety.

Policy: The executive champion and the senior management must insist that the product spec is released prior to authorizing the design and development phase.

Policy: The product specification must be released to the next phase ahead of all other documentation for that phase.

This would include signatures by the cognizant engineer and the cognizant marketing representative. If they can't agree, the executive champion or senior management must resolve the issue(s). Some companies have the chief engineer and the marketing VP sign the product spec.

The product spec will of course change as the development proceeds, but not without the control and visibility needed by all concerned. Unlike our constitution, the product spec needs to be considered a "living document" and kept in tune with the latest agreements. Engineering and marketing agree to the change (sign the redlined spec) and CM inputs the new system code and rev level to PLM/ERP.

CONTRACT/PLANNING PHASE

All new products will be assigned a new product part number and coded X in the system(s) by CM. When assigning any new item part number, CM will input to PLM and/or ERP a system with code X.

Any new drawings created are under the cognizant engineer's control and may be changed at will—code and specs likewise. Changes should be rev X1, X2, etc. and done by the cognizant engineer.

During this phase engineering is generally funded for design and development and they produce a product proposal and specification. Engineering will generally produce and test a "breadboard" model(s) in this phase.

Part and assembly drawings may or may not be prepared and part numbers may or may not be assigned to new parts depending upon the conditions.

The bill of material may consist of only two part numbers—the top level PN and the product spec as a quantity "Ref" and unit of measure "Doc."

If a "similar product" BOM is being used in the contract/planning phase, then all the items expected to remain unchanged should be included at their respective revs and codes and other items deleted.

With customer/management approval to proceed to design and development, the product spec will be signed and released for design and development (Rev date and system code E).

DESIGN AND DEVELOPMENT PHASE

When the management (or customer) determines that the planning is complete and funds the project, the design and development phase begins. We will term the units produced as "prototypes." The number of prototypes produced will vary depending upon the company, cost of producing a prototype, need for testing, etc.

The pilot schedule is normally determined early in this phase if not done in the planning phase. A bill of material may be developed from scratch for a new product or from a BOM of a similar product.

A cognizant manufacturing engineer or industrial engineer should be immediately assigned to answer design engineering's questions about the planned production processes and to assure manufacturability of the designs—attention executive champion.

The product specification should be pilot released (rev numeric and S code) during this phase. No other document should be pilot released until the product specification is pilot released.

All other drawings and specs must be pilot released in lead time to meet the pilot schedule. As released, CM will give them a rev numeric and S code.

The form used for release can logically be the same form as is used for changes, since most of the data needed is also needed for changes. It can also be a separate form. A blanket release form can also be used for all releases to pilot—see the EDC Handbook for details.

Policy: Only one engineering signature and the primary internal customer signature (ME or IE) on new design documents should be required for pilot release.

It should be incumbent on the ME/IE to coordinate with test, supply chain, service, etc. as needed. These other functions would naturally be involved if a release team is in place—and it should be.

It should be incumbent on the cognizant design engineer to consult with the other design engineers who might be involved.

When all items and the top level documents have been pilot released the senior management or customer should be appraised of all design and development issues, test results, etc. and make the decision to proceed to pilot or not.

PILOT PHASE

If designs are taken directly from development/prototype into production, the early units in production typically experience design and/or process problems that would have been caught in a pilot run. Since parts are now being purchased, produced, and assembled in production quantities, the cost of correcting those problems now may be 10 times as expensive as they should have been—"The rule of tens."

Policy: A pilot production phase will be utilized in order to minimize company risk.

Most new products beg for a number of units to be produced and tested either in-house and/or by friendly customers. Less expensive products may have dozens of prototype units built. Very expensive products may use the first unit to be built by production as the pilot/prototype.

The ideal pilot operation in this analyst's opinion is a separate function to buy, build, test, write fabrication and assembly processes, design the production facility, plan the materials issues, plan service and support, etc. This function should answer to the operations VP. The cognizant engineer(s) would also be physically with this group as should a CM representative. A pilot production manager should be assigned.

Regular team meetings of all affected parties would be held—at least once a week—face to face. While we want all affected parties to be involved in the pilot team, they need not sign design documents since they will have the ability to take exception to any release. If the team is working properly, they will have adequate notice to take exception.

Pilot rules are simpler than production rules but they put the product and its documentation under minimum control. Any changes to pilot released items should be accomplished by a separate change order—next numeric rev and still S code. Production control must be involved to furnish the effective date information. Normally all pilot units will be affected since changes in this phase are usually to meet the product spec—but PC needs to verify and implement this for every change.

If all pilot units are to have the change incorporated, the part number (tab) need not change even if the change is noninterchangeable.

The production schedule is typically determined in this phase. The product spec must be the first document to be production released—CM changes the rev to alpha and the system code to P.

The release of pilot items to production must be done in lead time. CM will change the revision level to alpha and the system code to P. No design or document changes should be made in the release—they should be done by change order prior to release.

Again, the author and acceptor must sign the documents for release to production—**in lead time per the production schedule**. Another approval may be added—perhaps a service engineer.

A separate form for each action or a blanket release can be used—see the EDC Handbook for details.

The entire bill of material should now be coded P and all documents at Alpha revision level.

When all items are production released, the testing of pilot units has been successful and customers are satisfied, the senior management will authorize the product released to production.

PRODUCTION PHASE

Begin the process of making profits. Thus when changes are "requested" or "required" in the production phase they should be:

• Scrutinized in order to screen out the unnecessary.
• Initiated by markup/redlining the affected document(s).

- Reviewed by all functions which might be affected.
- Authorized (signed) by very few functions/people.
- Incorporated into the master drawings and specs.
- Planned to be effective at an optimum point/date.
- Input into the PLM/ERP.
- Tracked/traced/status accounted.

Scrutiny will be discussed in the request process. Also see the change process for further detail.

OBSOLETE

Quite simply, when engineering decides that an item should no longer be used for new designs, they should write a change to obsolete that item. Engineering should decide if the item should be changed in some or all or no previous products using the item.

The change made on the drawing or spec would state: "Obsolete for new designs." The item document should be rev level OBS and coded O in the systems. See the EDC Handbook for further action.

SOFTWARE RELEASE

See the discussion in Chapter 4. Software engineering should operate under identical phases and rules as product engineering.

A printing equipment company had different phases and related coding in software engineering that in the rest of design engineering—wow—mass confusion. After the phases of release were agreed upon there was relative harmony.

As previously mentioned, the product CM folks need to be involved when software is transmitted to a customer and when a software release is made to pilot production and for all subsequent changes ("releases" in most of the IT world).

The initial product software release needs to be documented by the following:

- Spec with a part number—preferably dash 01—containing the following:
  - Version number of the software code
  - Applicable used-on product PNs
  - Build environment, tools, settings, and other pertinent data to allow regeneration of the media.
- Two copies of the media—one for production and one for a CM file marked as follows:
  - Part number (same as spec PN above)
  - Name
  - Version number

The object is to allow reproduction of the software just as we need reproduction of a part or assembly.

RELEASE PROCESS STANDARDS

The release process should be defined by at least the following standards:

- **Phase Release Chart and Policy**—Requirements such as those mentioned above.
- **Teams in Release**—Membership, chair, action items list, responsibilities, etc.
- **Release Form and Form Instruction**—The change form must allow for release of an item needed in a design change so the same form may be used for both. There must be specific instructions for completing the form for release—box by box—if online cursor pop-up instructions.
- **Release Work Flow**—Delineates the release procedure in work flow diagram form—online or not.

RELEASE METRICS

The release to pilot and to produce in volume necessarily needs to be done rapidly, it is critical to capturing the maximum market share. The time should be measured and reported.

As critical as the time to market is, it is somewhat difficult to measure. Of course key dates can and should be measured for each release of a document, an item, an assembly- or top-level product. The average time to release should be graphed as in Figure 8.5.

Continuous improvement should be expected unless a very unique product line is being introduced.

Perhaps more important would be a measurement of the time to release a new product. Figure 8.6 is an example for this writer's fictitious front end loader (FEL) company wherein a "baseline" from a similar product was previously established.

The data may not be fully comparable because the amount of new design effort in each product is not the same. Also the changes required to the new design after release are not taken into account. However, we are looking for trends and the trend, in this example, is positive.

To be sophisticated with this metric, we might add a line to the chart to indicate the percentage of new/unique part numbers in the product. This would allow us to put the time to release in perspective.

Also important to this metric would be a measurement of the number of changes to each new product in the first months of its existence. This would tell us whether or not we hurried up to do it wrong—a **process quality indicator**.

Thus, over time, tracking of and reporting changes to the design documentation might be in order.

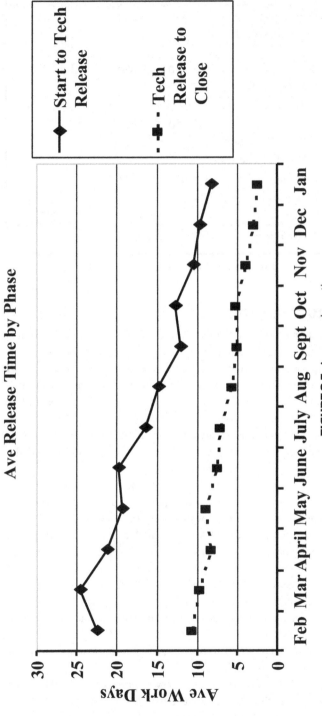

FIGURE 8.5 Average release time.

Phase	Description	Base	FEL-100	FEL-200
I	From Plan complete, to All Docs Released for Pilot	4.1 Mo	3.5 Mo	3.0 Mo
II	From All Docs Released for Pilot to Pilot Units Completed	4.3 Mo	3.1 Mo	2.2 Mo
III	From Pilot Units Complete to All Docs Released to Production	2.3 Mo	2.1 Mo	
IV	From Production Release, to First Production Unit	3.5 Mo	2.2 Mo	
	Time to Market	14.2	10.9	

FIGURE 8.6 Time to market.

RELEASE PROCESS QUALITY

Since engineering's product is documentation, it follows that we need to measure the quality of the documentation resulting from the release process.

Here are two measurements of the release process quality. The first is a short-term measurement of the number of changes (revs) to newly released documents in the first period of use—in this case the first 6 months. It can be prepared for a given new product or for all new documents released. See Figure 8.7.

We can see that the trend was negative for the first 5 weeks of tracking and then turned to a positive trend and then leveled out. Investigation is needed to understand what has occurred to prevent further improvement.

The ideal of course would be **no** changes to any newly released documents because the team was so good and the process so foolproof—but that is not realistic. Further improvement over time, however, should be expected.

If the team is functioning well and lessons learned from each product release are passed along to the next product team, the trend should be toward very few changes on new documents.

A longer-term measurement would relate to the changes required in each new product release over time. This metric is useful for both time to market and for release quality measurement. See Figure 8.8 showing the data gathered and Figure 8.9 graphing the results.

Notice that the last two products have not been in production for a year and thus the change data is not yet available.

We can analyze this data and learn some very important lessons. For example, notice especially that the FEL-200 was developed and released in record time but also with record changes in the first year of production. This probably

Release Process Quality Indicator

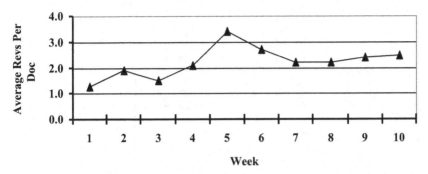

FIGURE 8.7 Average revs per new doc, last 6 months.

Product Type:	FEL	0	100	101	150	200	202	210	220
Time To Market - Mos		14.2	10.9	11.3	10.2	8.9	10.4	9.1	8.0
% of Unique PNs		42	27	26	19	28	16	22	20
# of Changes in 1st Yr		87	63	54	37	83	28		

FIGURE 8.8 Release time and quality.

required more service effort and possibly more field changes. If this was a conscious decision, the service was fast and done with little interruption to the customers, then goals were met. However, if this was a "hurry up and do it wrong" with customers' anger, then the record release time was counterproductive.

RELEASE PROCESS FLOW

For a best-in-class work flow diagram of the Release Process see Figure 8.10.

Note that events are put in parallel whenever possible. This makes for the fastest process possible. Notice also that the release is not considered closed until:

- The input to MRP, ERP, and PLM is complete and checked.
- The fabrication processes, publications, and assembly processes are complete as applicable.
- The CAD design files and code files are backed up.

Your organization should use this work flow as a guide in developing your own work flow. The key factor is to separate the design events from the administrative and operations event at the technical release event. Ideally CM

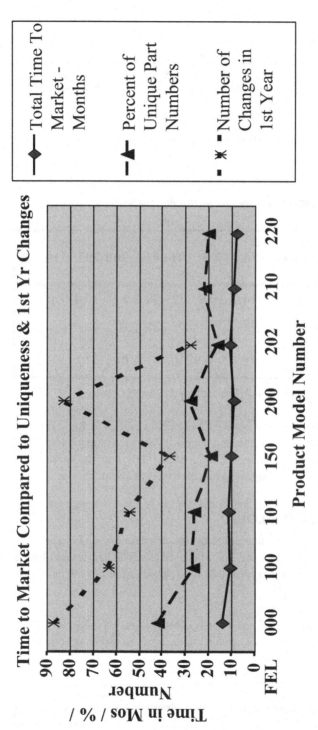

FIGURE 8.9 Release time and quality graphed.

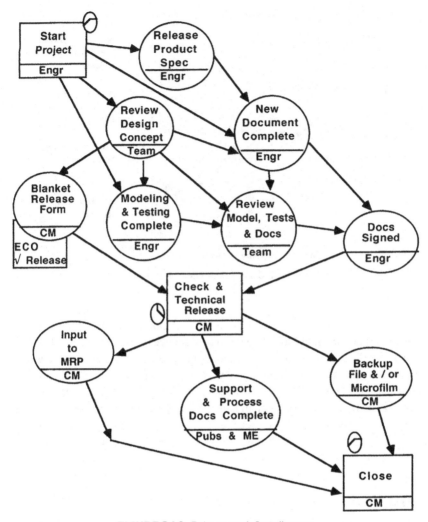

FIGURE 8.10 Release work flow diagram.

should chair the team and distribute an action items list immediately after each meeting—"minutes" are not necessary.

The events in boxes (with a "clock" attached) are the events which this analyst would time measure for each release whether by separate form/actions or logged into a blanket release form.

RELEASE TIME AND VOLUME MEASUREMENT

Measurement of each release action will provide the data to measure the time to release and volume of releases. One seminar attendee, who had a similar work flow, offered the graph in Figure 8.11.

This metric shows that while the average number of documents per release is going down slightly, the time to release is steadily improving at an even faster rate.

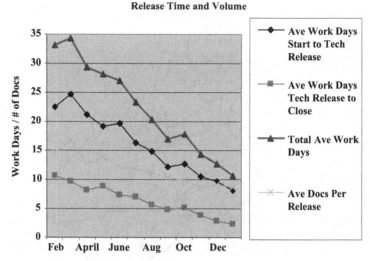

FIGURE 8.11 Release time and volume.

The establishment of a release "in lead time" policy was the primary reason for the improvement at this company. Their change system was not a deterrent to release. This analyst cares much more about releasing in lead time than the number of releases. If someone says, "we have too many releases," tell them it is the release lapsed time and release in lead time that is important, not the quantity of releases. Batching releases is, in fact, generally counterproductive—batching not to be confused with blanket release.

In any event, the trend in this case is very positive and some horn tooting was in order.

SUMMARY

For releasing a part, drawing, specification, assembly, or product from one phase to the next:

- Determine what the optimum phases are for your company—name and define them.
- Prepare a phase release chart for your phases.
- Develop rules for release to accompany the chart.
- Establish a streamlined work flow diagram for the process.
- Assure that standards are written for the process.
- Review/establish measurements and metrics for this process.
- Include facc-to-face team meetings in the process with all involved and committed (chickens and pigs).
- Allow for one author and one acceptor to sign each document.
- Require item release in lead time.

Chapter Nine

Bill of Material Process

Referring to the bill of material (BOM) as a "process" is perhaps misleading. The parts lists which make up an assembly (BOM) are released as part of the release process, changes are requested as part of the request process and changes are made as part of the change process. However, there are many aspects of parts list and bills of material that need to be addressed separately.

Each assembly must have a related parts list, the product BOM is simply a compilation of those parts lists—a simple definition for such an important document.

All assemblies need not have a pictorial drawing although a sound practice is to create a "no pictorial" document to clarify those which have and do not have a pictorial drawing.

After the product specification, the bill of material is probably the company's most important design document. It is certainly needed by more people in the company than any other document. Therefore,

Policy: It is company policy to assure that the design data portion of the bill of material is always 100% accurate.

This is simply a necessity in order to get the right parts to the right place at the right time.

So what is the design data in a parts list?

DESIGN DATA

The design data is listed in Figure 9.1.

The effective date (date-in and date-out) should be furnished by operations and could well be included in the above since the design data changes at some point in time.

Software and firmware to be embedded into the product is treated as a part or component. The part number and description should be listed on the associated assembly or product parts list.

Note that *nondesign* documents are *not* design data and will *not* be in this writers' BOM. If a bill of manufacturing documents or a bill of quality documents is necessary, they should be separately created and maintained by manufacturing or quality folks. This analyst would rather include them in the production and quality process documentation as they are an integral part of the manufacturing process.

Configuration Management for Senior Managers. http://dx.doi.org/10.1016/B978-0-12-802382-2.00009-X

Data Element	Comment
Part number	Primary / key data element
Document number	00 tab of every part number–represents the document
Description	Per standard–noun name, modifier, value
Cognizant design engineer	Relates to the document and the item.
Type of document	Per standard–assembly, part, doc, PL, etc.
Status code	XESP code indicating the valid phase
Size of document	Per standard–A, B, or C
Item weight	May not be required for parts but might be required for the product
Unit of measure	Only one per PN (must be the same regardless of used-on)
Assembly(s) used-on	Obtained via parts list entry

FIGURE 9.1 Design data defined.

NEW PRODUCT BOM

Design engineering produces a parts list with the assembly—often in CAD as a product of designing parts and designing an assembly pictorial. Engineering submits it to CM for entry into the PLM system.

A BOM is sometimes made for the new product by marking up a similar product BOM for CM to enter into the PLM/ERP system—new items being designed are entered with a new part number. Of course, where existing parts or assemblies are unchanged they may be left "as is" in the new product BOM, including rev level and status coding.

> **Policy: The parts list/bill of material for a new or spin-off product should be released to CM for entry into the PLM/ERP systems in the development phase.**

The release status of the **entire product and each item their-in can be easily seen by viewing a BOM showing the XESP coding.** This tool is crucial to managing and monitoring the progress of the development process.

CM should also assure that the same BOM is in the PLM and the ERP system. The most obvious reason for this for engineering is that the ERP may be the only way to develop and track the product cost.

At this stage, the BOM should be totally flat or minimally structured. Put off the structuring of the BOM until a consensus can be reached but no later than the product release to production.

PRODUCT COST

The ability to track the cost of the new product during the release process is obviously critical. This is true especially in the development phase.

A BOM is available in most ERP systems showing the cost of each item—standard or estimated. Most ERP systems have been designed to "roll-up" part and assembly cost to yield a product cost. That requires

- The purchased item cost,
- The machined or fabricated part labor and overhead rate,
- The assembly labor and overhead rate, and
- The ERP application to calculate the cost at part, assembly or product levels.

When the structure is not yet established, the assembly labor can be included under the end item part number. Thus, we can judge whether or not product cost goals are being met whether or not the structure of the BOM has been agreed upon.

Some PLM systems also allow entry and roll-up of product costs. You should certainly not want to maintain costs in two systems—another challenge for the executive champion.

Policy: The product cost shall be maintained in only one system—and autoloaded to the other if absolutely required.

The single system should normally be the ERP since several functions in that system require product cost information. Many considerations are involved—not the least of which is the relationship of the items cost to the purchase order, purchase decision reports, and price variance reporting.

PLM AND ERP

ERP and PLM systems normally include a BOM module. The ERP is designed for manipulating materials and operations issues and the PLM is designed for manipulating design documents.

The parts list data are often entered into the ERP system by operations folks. When they purchased the ERP system, they sat someone(s) down to do the data entry, often this was done before PLM systems were in wide use.

The PLM system allows easy retrieval of design graphics and specifications. They also have more or less robust work flow ability. It is by these work flow features that many companies handle their CM processes. ERP systems typically have no work flow ability.

Almost all ERP systems allow for both an engineering BOM and a manufacturing BOM structure—with the same set of parts—we hope. Some multiplant operations enter, for whatever reasons, the design data separately at each plant.

The PLM and ERP systems can contain unique BOM structures—often an "engineering structure" and a "manufacturing structure."

So the gap between engineering (PLM) and operations (ERP) widens! The necessity for bridging the gap and keeping these two systems "in sync" should be obvious.

Policy: The CM organization shall be responsible for entering the design data in both ERP and PLM.

The design data portion of that data entry should be done by CM to assure 100% accuracy and to assure synchronization of data and changes.

Policy: The personnel which have been entering the design data into the ERP system should be transferred to CM.

A further problem/challenge stems from the fact that most companies do not have a single BOM—that is, a single data entry of the parts list data.

Making CM responsible for the data entry to both PLM and ERP is only the first step toward achieving a single data entry.

ONE BOM ENTRY

Many parts of the organization rely on the design data in the PLM and/or ERP. Other systems in publications, manufacturing engineering, field support, accounting, etc., have a need for the parts list data. They often enter that data again in their system.

Some of those needs are:

- ME/IE preparation of assembly processes.
- Pubs/service for prep of spare parts lists and/or illustrated parts lists.
- Pubs/service for prep of illustrated maintenance manuals.

These needs usually include the graphics for assemblies. The above folks should have access to pictorial info in CAD or PLM without the ability to change it.

As the saying goes; "parts is parts." Certainly a product is made up of parts but it seems that every function in the company wants to manipulate the parts differently for their own purposes. As a result, BOMs/parts lists are maintained not only in ERP and PLM but also in CAD, publications, process/routing modules, service parts databases, etc.

Multiple data entry of BOM parts list data is not only a waste of data entry time but results in extra effort to reconcile those databases and to correct discrepancies. Or, left unreconciled, many material purchase, fabrication, stocking, and assembly errors are created and wasted efforts result and/or schedules are missed. Sometimes multiple BOMs allow bad design decisions to be made by someone other than the design engineering function.

The MRP/ERP and PDM/PLM systems usually do not talk to each other. Such duplicate entries of parts data remind this analyst's of his favorite industrial engineer story—being an ME by degree but an IE by nature:

A Plant Manager was looking for an Industrial Engineer (he had heard that they could save him some money) he asked each prospect to walk the assembly line with him and to point out any areas wherein the IE thought he could save the company money. As they walked the line, one IE noticed a person sitting and observing the assembly line. The IE asked the PM; "What does he do?" The PM said I'll find out—he left, asked, came back, and said; "Nothing!" The IE didn't say anything! As they went on down the line, the IE saw another person sitting and watching. The IE asked the PM; "What does that person do?" The PM went and asked and came back and said; "Nothing!"
*The IE responded very quickly; **"Ah Ha! Redundancy!"***

Multiple entries are all to frequent in industry. This analyst's survey of product manufacturers shows in Figure 9.2 just how pervasive the problem is.

The inefficiency of the redundancy of BOM data entry is only part of the story. It is easy to see that errors will result in any or all of those bills of material resulting in unnecessary expenses—wrong parts, missing parts, delays, confusion, time to correct, lost opportunity—effectiveness issues.

One of the most challenging opportunities in the CM world is to foster a single entry of design data. It is unlikely to occur without an executive management decision:

Policy: It shall be our goal to attain a "single BOM" by having entry of design data one time in one system for initial release and changes and allowing other systems to be downloaded automatically.

This should be done as part of the BOM process redesign or continuous improvement project.

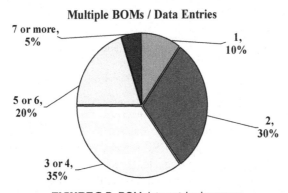

FIGURE 9.2 BOM data entries in survey.

The CAD system is usually the origin of the BOM generation. The CAD data is then furnished to CM by the cognizant engineer for their entry—usually via a spread sheet or sometimes downloaded.

The ideal situation would be for CM to download the CAD data to the PLM, add the necessary system coding and then autoload the ERP. This should be for new bills and changes. Some companies have systems that allow auto download of new and change data. If CM controls that download, it is another step toward a single BOM.

The ideal process would allow the pubs, ME, and field support engineer to access the ERP and/or PLM system to obtain the needed data **and pictorials** without having the ability to change the system.

The items to be "spared" should be coded in one system (probably PLM) by cooperation of the cognizant design engineer and service engineer with entry by CM. Alternately, a separate list of service items can be made as a reference document in the product BOM—see the spare parts list referenced in the engineering friendly parts list discussion below.

Such steps will allow the service parts folks to extract the item numbers and data to manage stocking of spare parts. Also the publications folks would be able to extract only those parts for listing in the illustrated parts catalog.

So much for parts—how those parts are to be structured is another, even more difficult question, which will be covered later.

Operations folks have, of course a need for many other data elements to be added to the BOM.

OPERATIONS BOM DATA

Manufacturing will input and maintain elements in the ERP system related to the design part number, such as:

- Make/buy code.
- Standard cost for purchased items.
- Lead time to buy or build.
- ERP codes.
- Hours to fabricate or machine parts made in-house
- Hours to assemble.
- Labor rates.

Although sometimes used by engineering, this data is not engineering authored and should therefore not be entered by CM. Manufacturing would enter that data via the appropriate ERP screens—purchasing, production control, manufacturing, cost accounting, etc.

Engineering will also enter data and links to the PLM systems for document search, file locations, etc.

ENGINEERING-FRIENDLY PARTS LIST

Bills of material/parts lists produced by the ERP and sometimes even the PLM system are often considered "unfriendly" by engineers and rightfully so.

The structures often aren't as they deem most logical. They wish to use a parts list for changes and the systems don't allow "redlining" as most CAD graphic systems do. They often contain nonengineering data that just clutters their vision.

Policy: An "engineering-friendly" parts list shall be programmed so that every assembly list shows only design data (plus date-in/date-out) and is double-spaced for easy markup.

This friendly list may be from either the PLM or ERP system as long as CM maintains the design data in both. It should be done as part of the BOM system reengineering or continuous improvement project.

The engineering-friendly parts list would look like Figure 9.3.

It contains only design document/part numbers as discussed. The next change can be readily redlined because of the double spacing feature. Many errors in BOM changes can be avoided by this simple method.

A medical device company adopted this approach and later reported to this analyst that they had nearly eliminated their change BOM data errors.

The product cost data might be included in this engineering-friendly parts list—although it is not design data per se—it might be well to keep it ever visible to the engineers and others involved in CM.

Notice it does **not** show the revision level on any of the **components** in the parts list. This is done purposely to avoid "rev rolling"—more discussion on this in the change process chapter.

Best practice is to use the same PN on the assembly pictorial drawing as on the parts list. Another cross reference avoided.

If this assembly/product had embedded software, another line item would be included for the software specification PN as previously defined.

All changes to the BOM can be traced to the effective week-date by this feature. Example; the small tire was changed to the 02 version on week 48. Such "date tracking" is acceptable by most agencies and good practices. When a change needs to be tracked to the serial number (or equivalent) it is not done in the BOM—more about tracking in the change process.

BOM CONTENT

Almost every company has an ongoing debate about which items belong in the BOM/parts list and which don't. In short,

Policy: Any item that is part of the product when shipped as well as design documents defining those items or defining the product should be listed in the BOM.

DATE 7-12-99	REV B	DESCRIPTION Small Tire OD		ECO # 1212		SIGN FBW	
EC3 CORP FEL - 100		DESCR Final Assembly		P / N 223456-01	SIZE A	PG 1	OF 1
FIND #	DESCRIPTION		PART NUMBER	QTY	UNIT MEAS	IN/OUT DATE	
1	Motor Mount Asm		223356-01	1	ea		
2	Tire, Large		423456-01	2	ea		
3	Frame Asm		723456-01	1	ea		
4	Tire, Small		423456-02	2	ea	wk 48	
5	Bucket, 4 yard		523456-01	1	ea		
6	Bucket Arm		823456-01	2	ea		
7	PCB, Elect Ign Asm		923456-08	1	ea		
8	Nameplate		323456-01	1	ea		
9	Axle		103456-01	6	in		
-	Product Spec		123456-00	Ref	Doc		
-	Material Spec		623456-00	Ref	Doc		
10	Wheel Hub, Large		113456-01	2	ea		
11	Wheel Hub, Small		121456-01	2	ea		
12	Motor Asm		114456-07	1	ea		
13	Adhesive		115456-01	2	oz		
-	Spare Parts List		623457-00	Ref	Doc		

FIGURE 9.3 Engineering-friendly parts list.

All items that represent engineering's product design responsibility are included. This would include labels and nameplates. It would not include manufacturing or quality documents. It would include all floor stock items although they need not be quantified—quantity "AR (as required)" is acceptable.

In addition, some companies may want to include the material used to package the product and/or the manuals shipped with the product.

Whether or not packaging items are included in the BOM, we should investigate how packaging is handled. Is there a clear responsibility for packaging design? Many a new product has reached the shipping floor only to sit there for lack of packaging.

Some folks want to include all the ME, QE, service or sales documents related to the product. Those documents should be included in the process documentation done by those functions or, less desirably, in a separate BOM—not in the product design BOM—and not controlled by CM.

STRUCTURING THE BOM

Considerable time and energy is often wasted in early debates about the product assembly structure. Debate about the BOM structure should be put off for resolution near the end of the pilot phase.

A single-level structure or a partially structured BOM should be adequate for pilot production purposes. What is/is not to be an assembly is normally not important to the pilot release or the pilot build process.

The design release team should address this issue with an eye toward minimizing the levels of assembly in the BOM. Assembly pictorials are a manufacturing tool. The creation of assembly part numbers and pictorials may be largely wasted, if manufacturing processing is done with downloaded CAD pictorials or photographs.

The service assemblies can be created and made part of the service parts list. Engineering's need for assemblies (standard to more than one product) usually matches the servicing needs for service assemblies.

More levels beget more complexity in every BOM change, where used search, and most important the temptation to stock assemblies is great. The driving force behind JIT is to eliminate stocking of in-process assemblies—thus eliminating the very high inventory carrying costs. Inventory carrying costs are often estimated to be as high as 65% of product cost. In fact, companies who use just-in-time (JIT) techniques often only have single-level structures. In most companies, a minimal/shallow structure is attainable.

EVOLVE THE BOM

Regardless of the depth of the BOM, the structure should **evolve over time**:

- In the planning phase, a top level part number and the product specification document number are all that is needed in most environments.
- In the design and development phase, the items needed should be put into the BOM under the top level part number without further structure.
- During the pilot phase, the items should be released in lead time and minimum structure added as agreed upon.
- Before release to production, CM should "put away" all items into the mutually agreed upon assembly structure.

Pictorially that evolution looks like Figure 9.4.

Many structuring issues need to be faced if we are serious about bridging the gap between engineering and the rest of the operations by attaining a single company structure. Issues such as:

- Multiple plant build with different tooling or methods.
- Stocking an item.

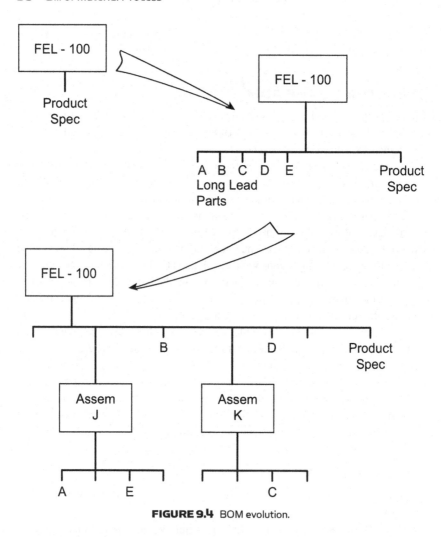

FIGURE 9.4 BOM evolution.

- Buy an item from more than one supplier (fabricate part/finish part, for example).
- Shipping between buildings.
- Accounting reports for line managers.
- Firmware/application software.
- Standard assembly.
- Spared assembly.

Some form of identification is needed for shipping between plants. Firmware and software structuring will be discussed shortly—the remainder of these historic "reasons" for added structure levels are discussed and refuted in the EDC Handbook.

The minimal structure should be agreed upon between engineering and operations. Other organizations do not have a significant stake in the structure and usually need not be involved.

Many folks say that agreement is impossible or far too time consuming to even try. They haven't tried. They point to the two BOMs available in all ERP systems and say—see, they know it can't be done. The designers of PLM and ERP haven't tried either.

Some say as long as the same parts are used, what difference does it make? Operations can create "phantom assemblies" and have done that for decades. The same folks are quick to say that engineering throws the documentation over the wall or that manufacturing doesn't know how they are going to process the product. Has anyone ever figured out how many hours are spent in interpreting the differences, translating the communications between parties, assuring that the same parts are used, calculated the cost of errors resulting—throughout the product life cycle—forever?

The operations folks need to get involved earlier, the engineering folks need to avoid structuring too early and make it a goal to develop a structure that will work for both organizations. Yes, the first time it is done it will be painful and time consuming. However, shall we continue with the gap between engineering and operations forever?

One of the keys to success is to flatten the structure—fewer levels.

JUST-IN-TIME BOM

Many companies have shifted to just-in-time (JIT) manufacturing with a single-level BOM. One computer manufacturer does it with a product of a 1000 parts and components. That technique might well be examined for application in your company.

STRUCTURING FIRMWARE AND SOFTWARE

The usual method for structuring firmware is to make an assembly out of the unburned chip and the program. The combination is given a distinct part number as follows:

PN	Description	Qty/UM
Z	FW (burned chip) assembly	1 ea
Y	Unburned chip	1 ea
X	Burn program	Ref doc

The burn program would be called a ref-doc in the quantity-unit of measure fields. Thus, the system would not drive a burn program for every unit to be produced. This method makes an assembly out of X and Y. This adds a level to the structure, however.

Is this the only way to handle firmware? No. An alternative that eliminates this assembly is to structure in the next higher assembly or in the product assembly the burn program and the assembly with a reference designator similar to that used in printed circuit board design.

For example, let us name this particular chip the QPL function. Then the product can include in its parts list a reference to that designator (QPL) in the description of the program and the assembly. For example:

PN	Description	Qty/UM
X	Burn program for QPL function	Ref doc
Y	Unburned chip	1 ea
Z	Burned chip QPL	1 ea

As long as this is a one-for-one relationship it works. The test group (or whoever burns your chips) will understand how to program the chip.

If other chips are to be burned from the same chip part number, Y, this method still works merely by increasing the quantity of Y. However, if the same unit contains more than one type of unburned chip, this method then breaks down.

Applications software can be handled similarly by giving the program and the programmed media a reference designator.

FEATURES AND OPTIONS

Often products are offered with a myriad of features and options. There are various ways of handling this situation:

- Order-related bill of material
- Customer configured order
- Modular BOM
- Configurator module purchase

These alternatives are thoroughly discussed in the EDC Handbook, especially the modular BOM concept. They each have circumstances that make them a desirable choice. Before purchasing a configurator module, the use of the other alternatives should be investigated.

THE PERFECT BOM

Although the BOM may be different for different companies, there are some attributes that should be common. Let's summarize the 11 most significant attributes:

1. One data entry.
2. Must be 100% accurate, at least with regard to design engineering data.
3. Contains part, assembly, code, and document numbers defining those items which are part of product, and no more. (With the possible addition of the product packaging or publications shipped with the product.)

4. Design engineering data are input by CM, manufacturing data by manufacturing, materials data by supply chain, etc.
5. Is feature and option modular if the product lends itself to that concept.
6. Has one level (if JIT) or two levels (if feature and option modular), and no more than three or four total levels.
7. Contains the database elements (defined in a dictionary) for design, manufacturing, field service, and accounting.
8. Has date effectivity ability and historical record of same.
9. Can display the used-on assembly part number(s) and the corresponding used-on product part numbers/model numbers, preferable in one look-up/screen.
10. Will produce a variety of reports on demand. One of these reports must be a double-spaced engineering friendly parts list. Product, assembly, and part cost reports are also critical.
11. The single, shallow structure that has been jointly developed by engineering and manufacturing.

These are only the attributes most important from a CM standpoint. There are other criteria that manufacturing, accounting, or field service, would add to the list.

The inclusion of product cost data is critical to the company management. This list is a practical guide for companies considering purchase of an ERP system. There are other attributes that are important to find in a PLM system.

BOM WORK FLOW

Diagramming the BOM work flow is integrated in the release, request, and change work flows covered in those chapters.

The initial release and update of the BOM design data linking CAD, PLM, and ERP databases, with one data entry, is easier said than done—ideally;

1. Parts list data are developed in CAD as part of the item design.
2. CAD parts list downloaded to PLM by CM.
3. CM adds structure, rev level, and status code
4. CM downloads BOM to ERP
5. Manufacturing data added to ERP
6. Cost, effective dates, and other data as necessary are uploaded to PLM (can be eliminated if the ERP is available to engineers) and they are trained to use it.

It is probably obvious to even the casual observer that the ideal situation would be for **one system that does the job of both PLM and ERP**—don't hold your breath.

BOM STANDARDS

Several general standards or other process standards relate directly to the BOM, such as:

- Part numbers
- Quantity and units of measure
- Bills of material
- Approved manufacturers list
- Part number and revision level changes
- Effectivity
- Effectivity management

BOM METRICS

The most important aspect of a bill of material is its correctness of the design data and the redundancy of data entry which multiplies error potential.

Certain errors in the BOM should be found by configuration management during pilot release. For example, the engineer specified a part with four mounting holes but only specified three sets of hardware. CM technicians should see this discrepancy and correct it before entering that data—talking to the engineer if any doubt exists.

This is part of the quality control aspect of CM. CM will also make data entry errors which need to be corrected. Each line item corrected should be counted and reported, partly for horn tooting purposes.

In another case an error may have been made on previous entry of design data for a release or change. Such issues, when found, should be quickly corrected by a change order. Correction changes should be counted and reported.

If more than one data entry exists in your company, the reconciliation (planned or accidental) will find differences in the design data. Correction should be done via change order and counted. This is easier said than done since CM will have to gather the data from all who do their own data entry.

These correction and reconciliation actions should be reported to top management. The report might look like Figure 9.5.

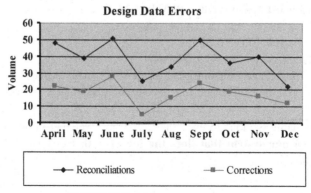

FIGURE 9.5 BOM design data errors.

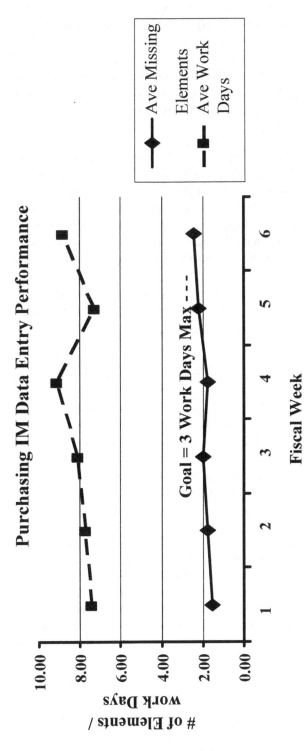

FIGURE 9.6 Item master data entry performance.

The goal, of course would be to see both measurements go to zero. Reduction of the number of databases (data entries) will be the most significant step toward reduction of reconciliation errors.

CM is not the only organization that enters data to the ERP system. The ERP item master file is normally populated by various departments. Each organization entering data to the ERP should measure the time from CM design data entry until they have completed the required elements on their screen. An example for purchasing is shown in Figure 9.6.

In this case the purchasing people have about two missing data elements in any given week but it takes about eight work days to enter the missing elements. A reasonable time was determined to be three work days. A computer peripheral company had this issue and the delay was diagnosed to be "waiting for supplier quotes". This meant that all the part, assembly and product cost roll-ups were eight days lagging. We designated a seasoned industrial engineer to set the purchased price standard and eliminated the delay. The standard was better than quotes because it gave purchasing a bogie to aim for, rather than using the first supplier's first quote.

The exec champion should assure that these measurements are done and that a goal is set for each department required to add data to the item master file.

SUMMARY

The many aspects of the bill of material "process" need to be understood and separately addressed:

- Need for design data to be defined and 100% accurate.
- A single data entry for design data is highly desirable for effective and efficient bills of material.
- What is and is not included in your BOM needs to be defined.
- Product cost data must be included in the BOM.
- An engineering friendly parts list should be available for markup of changes.
- Minimum BOM structure levels are needed to reduce inventory carrying costs and help bridge the gap between engineering and operations.
- All functions that input data should measure and report accuracy and timeliness of completion.
- Assure that metrics are in place and that related standards are written.
- Evolve the BOM as the items are released in lead time—both for pilot and production.

Chapter Ten

Request for Change Process

Some operations do not have a standard method of requesting document or design changes. Some do—called by different names—engineering action request (EAR), change request (CR), etc. Some companies use the change form for requests.

Requests for document or design changes can come from marketing, service, customers, suppliers, field failure reports, etc. A conscious decision should be made as to how each will be processed. This discussion will cover the internal company change request process no matter who uses it. They might well all be treated the same.

A few companies call for a "design freeze." This is an unrealistic attempt to reduce the number of changes. Better to place the product on the "do not improve" list and still encourage nonimprovement changes. Even automotive "model year design methods" have changes during the year.

It is best practice according to this writer to separate the request process from the change process.

SEPARATE REQUEST PROCESS

Some companies have a statement on their change form or in the change standard, which claims that "any employee can originate a change." They are probably using the same form for request and change.

This statement is usually patently untrue and if truly meant—stupid! Many employees who experience document-related problems do not know how to originate a change to the drawings or specifications. They don't know how to fill out a change form to make a request. They often don't even know how to find a change form. So they mumble and try to remember to tell the IE/ME about their problem. Thus, many issues go unresolved and are managed by added labor or by flatly ignoring the design documentation—"redline it and let's move on!"

Also, does that "any employee" statement really mean that every change request will be treated as "an automatic need to change"? Those folks who use the change form to request change are typically in this fantasy world. This "leap of faith" is a major contributor to the backlog of "change orders," which exists in most operations. They don't have the engineering manpower in the next decade to address all of the requests that pile up.

103

Configuration Management for Senior Managers. http://dx.doi.org/10.1016/B978-0-12-802382-2.00010-6

In very small operations, the request process is a shout across the garage. Some small companies accept an email to the CM person as the Request. However best practice;

> **Policy: All but the smallest of operations need a separate process to allow operations, service, and other folks to request changes to design documents and product design.**

Having a means to "request" does not mean a free pass for incorporation of every idea. All ideas are not necessarily good ones. Care also needs to be taken to avoid the "suggestion box" syndrome—wherein the requester never hears back or the idea is rejected only to find some months later that their idea is being implemented.

DOES YOUR REQUEST PROCESS WORK?

If you have a request process and need to determine if it works, the executive champion should take a tour through industrial engineering, the assembly floor, receiving inspection, fabrication, machine shop, purchasing etc., and ask a couple of people these two questions:

1. "Have you requested engineering to make any document or design changes? If the answer is "no" move on. If the answer is "yes" ask;
2. Have they responded to let you know whether or not they will address your issue?

This consultant uses this method when asked to do a process analysis. It will not take long to find out if your current process works. You may have to brace yourself for the blunt statements that often follow. They may have asked long ago and haven't heard a peep.

On the other hand, you need to be ready to handle the requests openly and quickly.

COST REDUCTION AND IMPROVEMENTS

Far too many cost reductions **aren't**, and many products that don't need to be improved **are!**

One electronics company had 11 "continuing engineers" processing changes. Analysis showed that an equivalent of four of them were processing requests/ changes that were called "cost reduction," "save time," "ease-of," etc.—which didn't have payback. When a payback standard was set and a sampling of their history was analyzed—a new policy was set:

> **Policy: Requests for changes to save labor or reduce costs must be cost analyzed and must payback the one time costs involved within X months.**

You define "X" for your company. Some companies may expect payback in a few months and others within a few years. As a guide, look at your policy for payback justification of purchase of new capital equipment. The cost calculation and payback methods will be addressed in the following chapter.

In the company mentioned, one of these people was used to calculate the payback of requests—most of which didn't meet their new payback standard. Thus, by defining payback and estimating the cost correctly, they had three engineers "in excess."

That person also worked with management to develop a "products not to improve" list—which saved an equivalent of two more engineers. A total of five engineers were returned to development engineering—a windfall of badly needed people.

> **Policy: Executive management needs to make a conscious decision as to which products need to be improved and which ones don't.**

The word "improve" is used purposely to mean above and beyond customer specifications. Every company that has products which have been more than a few years on the market, should put all products into either "improve" or "no improvement" list. Or code their products in the PLM system. Those lists should be reviewed every 6 months or certainly every year—executive champion and CM again.

SCREENING REQUESTS

Most request processes do not make it clear as to whether or not and when the engineering organization takes ownership of requests. The result is a pile of requests laying in engineering but for which engineering doesn't feel responsible—and the requester thinks they are.

One truck manufacturer had over 2000 such requests on file online. They had lots of frustration, anger, and redlines on the other side of the bridge. The operations folks were essentially "doing stuff as they saw fit!"

When this writer did design work, there was a pile of requests sitting in front of him that gave him a feeling of comfort about his job. He knew that he couldn't possibly satisfy them in the next year or two; wasn't sure they should all be done; and didn't understand some and didn't have solutions for some. Is letting requests pile up like that the best process?

An analysis of your requests or changes may be in order. One company did such an analysis and got information as shown in Figure 10.1.

In this example, 23% of the requests should be rejected by policy—the 14% and 9%. Also, 48% need to be analyzed to figure out if they are necessary and if manning is available to address them in a reasonable period of time.

Engineering does not have unlimited manpower and we need to quit treating requests as if they did! We need to have an effective review of most requests and methods for rejecting many of them.

Analysis of Requests/Changes

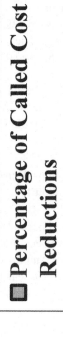

■ **Percentage of Called Cost Reductions**

■ **Percentage of Real Cost Reduction**

□ **Percentage of Improvements to Mature Products**

□ **Percentage of Improvements to Developing Products**

■ **Percentage of All Other**

FIGURE 10.1 Analysis of requests.

Policy: Product manufacturing is not a politically correct world—an effective method of screening requests needs to be put in place.

The best practice this analyst has participated in was a two-phase approach to either "technically release" requests to the change process or to reject requests.
Phase I:

- The CM organization did an initial screen to:
  - Accept any request for correction of document errors, notify the requester of acceptance and, do a change order immediately.
  - Reject certain requests based upon rules developed in Phase II and notify the originator.
  - Move requests needed to make the product meet specs directly to the cognizant engineer and notify the requester.
  - Move requests that involve a "line-down" situation directly to the cognizant engineer and notify the requester.
  - Take the remainder of the requests to Phase II.

Note that CM can and should write the change order for document only changes.
Phase II:

- A team of VPs reviewed all the remaining requests to:
  - Briefly review the CM Phase I actions since the last meeting.
  - Reject all requests to improve products, which need not be improved. (As time went on they developed the "do not improve list" and delegated this task to CM.)
  - Reject all requests that engineering did not have the manpower to address in the next 6 months.
  - Reject any request that would prevent engineering from working on more important design work in the next 6 months.
  - Reject those requests judged not to payback per the standard set.
  - Ask engineering for a probable design fix and payback cost analysis on some requests.
  - Determined that some requests should be incorporated in the next product iteration. Chief engineer ownership, CM kept a list.
  - Move the remainder of the requests to the cognizant engineer—CM notifies the requester and starts the change process clock.

CM logged all requests, chaired the meetings, and published the action items list within 1 h of meeting adjournment. CM logged and notified the requester of rejection or acceptance. CM produced reports on volume and throughput time as shown later.
Their request process averaged four work days lapsed time.

REQUEST REVIEW TEAM

Engineers are reluctant to reject requests. They find it difficult to say "no." A high-level management team to reject requests is needed.

The best practice team is made up of the chief engineer, the VP of operations, and the VP of supply or their high-level representatives. They meet three times a week. They quickly develop rules of thumb that allow reviews to occur in half hour meetings.

Some companies prefer to have the cognizant engineer at the executive team meetings. The CM manager should chair the executive team.

> **Policy: A director or VP from engineering, supply chain, and operations shall review most requests that require a change of design.**

Does your company or division have unlimited engineering manpower?—probably not. This analyst has never witnessed such a condition. Thus, politically **incorrect** action—**rejection**—is often in order.

Such a team will have the sensitivity needed to evaluate the lost opportunity, payback, and available manpower to reject many requests.

They will also develop some rules regarding which products should/should not be improved, thus rejecting many so-called "improvements."

CM can and should process BOM, drawing, and specification error corrections without submitting them to the management team. You might not even require that the engineer be involved with such corrections—only notified.

Such a management team can also develop rules for the CM manager to both reject some requests and forward some directly to the engineering folks without the management team review.

> **Policy: The team will develop and the CM manager will write rules (standard) for the types of changes that need not be screened by the team and can be rejected or accepted by CM.**

> **Policy: Ditto for the types of changes that can be forwarded directly to the design engineer for action.**

This is the most practical method of reducing change order backlogs and to attain quick response.

The team may ask the CM manager to find out more information or to have the engineer and/or the requester attend the next meeting to add information.

The requester must be promptly notified of the disposition of their request. From the time of receipt of the request in CM until the requester is notified of the go/no-go decision, an **average of three to five work days** should be the goal in most companies—with current manual process and legacy systems.

The CM manager should keep a log of all requests and their status and disposition. That should include date and time received, action pending and the approved/rejected date and that log should be available to all who need to know—online.

Software code change requests should be handled by a similar high-level team. CM may or may not be in that loop—SCM should be if CM is not.

REQUEST WORK FLOW

Thus the work flow for the request process is fairly simple.

Anyone/Originate a Request Form

Need not have a solution—only a problem.
Sent directly to CM.

CM Does Initial Screen

→ (Enter on tracking log, reject some, do the change for doc only issues, meet spec to cognizant engineer, and send rest to executive team)

Executive Team Screens Per Rules Developed

→ (CM Manager rejects or accepts per team decisions, updates tracking log, clocks end of request and starts change clock on those accepted)

Cognizant Engineer

→ (Start change process)

Realize that in order for "anyone" to request a document or design change they need to have access to the system or a hard copy form to do that.

It is not wise to have requests submitted to the requester's manager, IE/ME, or any other third party before it is sent to CM—they will only inject delay or may change the intent.

Note that the acceptance or rejection of any request is fed back to the originator of each request as soon as possible. The request clock stops only when that is done.

Only the CM manager's signature should be required on accepted or rejected request forms. When a request is accepted by the team or CM, the clock should start on the change process.

REQUEST STANDARDS

There are at least three separate standards that need to accompany the request work flow system/diagram:

- **Change request policy**—policy statements including those mentioned above.
- **Team in request**—membership, chair, action items log, responsibilities, etc.
- **The request form**—A separate form is most desirable. A form instruction, whether online or hard copy, is required. More details in the *EDC Handbook*.

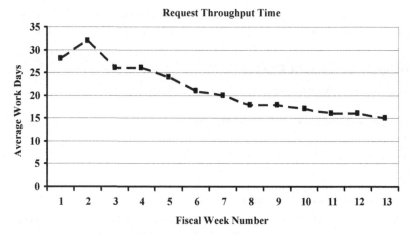

FIGURE 10.2 Total request time.

REQUEST METRICS

Perhaps the most important of the request process metrics is the time from origi-
nation of the request to acceptance or rejection. Either engineering will take
ownership of the problem or not. See Figure 10.2.

A goal for the through-put time should probably be added to the graph. It
need not be the ultimate goal but rather an interim goal with a date for expected
achievement.

The turnaround time in this case is, although improving, still deemed excessive. In order to analyze the causes, a breakdown of the total time may be necessary. See Figure 10.3.

The work week numbers in Figures 10.2 and 10.3 do not correspond—both
are independent examples.

From Figure 10.3, we can certainly ask many questions about the process
time; Why is 6–10 work days needed to get a request into the hands of CM?
Who is touching the request—the originators manager, the ME involved, etc.?
What do they do with it? Similar questions for the other segments should be
asked.

There is one other measurement that is definitely worthwhile for executive
management—see Figure 10.4.

In this example, the rise of the number of requests in work in process (WIP)
should be alarming. It means that the total time metric will be increasing as
the backlog is building. One should also be asking why the number of requests
accepted is going up and rejections down? Is rejection politically acceptable? Is
a high-level team needed?

FAILURE REPORTING

The importance of the feedback of failure data is obvious from watching the
news—about recalls which have obviously been far too-long-a-brewing.

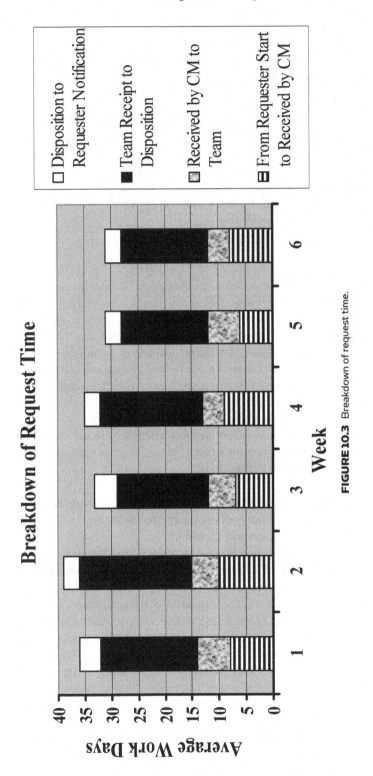

FIGURE 10.3 Breakdown of request time.

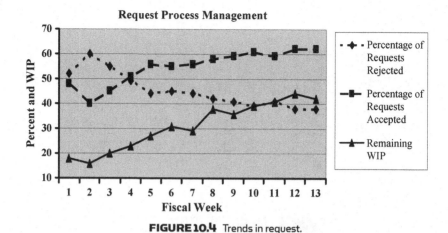

FIGURE 10.4 Trends in request.

Product failure information is critical to an engineer's ability to change the product to correct those design issues. CM usually doesn't get into this act, but it usually needs attention, so perhaps they should.

Failure reporting is one of the most difficult communications facing any manufacturing company. Feedback from the product test area is usually dependable. However, service people or dealers are performing repair and maintenance at the customer's site, often remote from the factory and from engineering. When product is returned for repair or refurbishment, the people in that process are often remote from the engineering function.

Even the operations test failures (product test, reliability test, etc.) are often not made available to the engineer on a timely basis. For these reasons, failure reporting needs to be carefully documented:

- **Policy: Practices regarding timely and useful failure reporting must be put in place and monitored regularly. Decide if this is a CM task or who else the process owner should be.**
- **Failure form—information needed for each failure for analysis.**
- **Report formats—specifies raw data and reduced data format(s).**

The policy should set a goal for timely feedback of failure data to engineering from all points of test and use. Failed items should accompany the data. It would seem reasonable to see this occur weekly with no more than a one week lag.

That data will usually reflect a "trial and error" repair process. When a failure occurs, often more than one part, component, or module may be replaced and "no problem found" on some items will result. Such condition requires analysis. Multiple failures require summation. The data reduction and analysis should occur in engineering.

The chief engineer should be intimately involved in this process. The first step is to determine what information is needed on the individual failure report

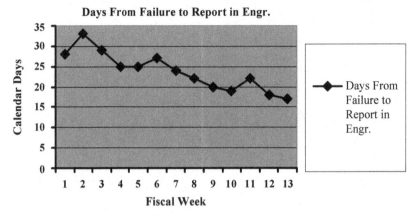

FIGURE 10.5 Failure reporting time.

and to design this "form." Then the time from the actual failure to arrival in Engineering should be measured. See Figure 10.5.

This report is based on feedback of the raw data—however that is done. Service folks need to assure that every failure gets reported.

One might also measure the time from arrival in engineering to change order or other disposition.

Executive champion—should CM be involved?

SUMMARY

To properly find and fix real documentation and design issues and to reduce the engineering change backlog, all but the smallest company needs:

- A change of attitude about the treatment of requests—**rejection is okay**.
- A well-designed request form and process—online *and* hard copy.
- The CM organization and executive champion to be the owner of the process.
- Rules for payback must be delineated.
- All products placed in either an "improve" or "don't improve" or "include in next product iteration" lists or coding.
- A very high-level team to review most design change requests in order to reject as many as good judgment and engineering manpower dictates.
- A simplistic work flow process under CM purview.
- Standards and metrics developed for the process.
- Clear determination of CM's roll in failure reporting.

Chapter Eleven

Change Cost/Payback

Bad habits can develop in start-up environments. Typically, the cost of a change is not an issue in a new product's early life, because essentially all changes are made to meet the product specification or to correct errors. It is usually a year or more before cost reductions and improvements enter the picture. That start-up situation fosters a bad habit—not to calculate the cost of any changes.

COST OF CHANGE

As we have already discussed, the so-called "cost reductions" often aren't—**if one-time costs are expected to be paid back in some reasonable period of time**. And, over time, all products do not merit improvement.

It follows then that not only does a policy need to be developed to specify the length of payback time, but also a method needs to be put in place to calculate payback when it is not obvious.

How much do changes cost? In the mid-1990s, a college professor decided to find out the cost of changes. He gave up because few companies calculated the cost, and those few that did included different cost elements.

When the author asks that question in seminars, the answer is typically a dollar figure between $1000 and $3500. Are these numbers meaningful? Are these the criteria that should be used to evaluate a change? Using this logic, if our company has a cost of $1500 per change, should we merely ask if the change seems worth $1500 or more?

Such numbers were usually developed by adding up the budget for Configuration Management and some other functions like revision drafting, BOM entry, and then dividing that number by the number of changes in the same period. The result is not the cost of a change—it is the administrative cost per change. This is not a bad number to have, however, if you benchmark other similar companies and carefully compare the functions included. This might be a good number to roughly estimate the cost of a "document only" change.

Every mature company should analyze changes (or better—the requests) to identify the potential for request denial, change volume reduction, and backlog reduction.

Combining the payback analysis with the "improve/don't improve" listing was shown in the example study in Figure 10.1.

Configuration Management for Senior Managers. http://dx.doi.org/10.1016/B978-0-12-802382-2.00011-8

That study showed:

- Called cost reductions but aren't = 14%
- Real $ reduction = 11%
- Improve mature prod = 9%
- Improve developing prod = 18%
- Others (meet specs, doc only affected, etc.) = 48%

The study at this company indicates that at least 23% of their requests should be rejected. Then the question of available manpower would cause rejection of another portion—although some might be good ideas for the next product iteration.

REAL COST REDUCTIONS

Changes that do reduce product costs (within the payback period) are worthy of tracking—and some horn tooting. Such cost reduction reporting should reflect only the direct material and labor reductions. And only those accepted and implemented that meet the payback policy. The cost reduction tracking should not reflect the implementation or start-up cost because real reductions in those areas are unlikely.

It must also be realized that the real cost reduction will not occur until the payback policy period has passed. Thus if the loader company has a one year payback policy, the real reduction in company bottom line will only be reflected one year after each real cost reduction.

Some companies set a goal for real product cost reduction. Care must be taken that material and labor cost *estimators* are detached form the ownership of the cost reduction goal. If not, it is too easy to attain goals with inflated numbers. This is not to erode the meaningful nature of the payback calculation but rather to caution against setting goals for cost reduction. The seemingly simple step of adding cost reduction goals will probably run the risk of number inflation, which can defeat the purpose of payback analysis.

One good metric measures and reports product by product, those changes that have had a payback calculation and were implemented, as in Figure 11.1.

Certain requests might be accepted by the request team and still be rejected by the change team after doing a cost calculation. The change team should not proceed if they have doubts about the payback. They should ask CM to gather the costs and calculate the payback.

Such policies and metrics are critical to the reduction of the number of changes and the maintenance of profits.

Policy: Avoid creeping elegance—calculate payback.

If you think calculating the cost of changes is expensive—try not costing them.

FEL-200 Unit Cost Reduction

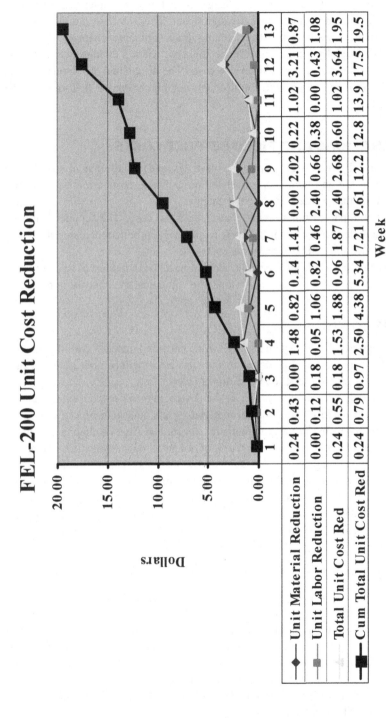

	1	2	3	4	5	6	7	8	9	10	11	12	13
Unit Material Reduction	0.24	0.43	0.00	1.48	0.82	0.14	1.41	0.00	2.02	0.22	1.02	3.21	0.87
Unit Labor Reduction	0.00	0.12	0.18	0.05	1.06	0.82	0.46	2.40	0.66	0.38	0.00	0.43	1.08
Total Unit Cost Red	0.24	0.55	0.18	1.53	1.88	0.96	1.87	2.40	2.68	0.60	1.02	3.64	1.95
Cum Total Unit Cost Red	0.24	0.79	0.97	2.50	4.38	5.34	7.21	9.61	12.2	12.8	13.9	17.5	19.5

FIGURE 11.1 FEL-200 cost reduction forecast—first quarter.

DESIGN AND DEVELOPMENT COSTS

The engineering time to analyze, design, model, test, document, and communicate the design change is usually significant. This is a cost that the engineer might mentally evaluate before launching into a significant change.

It is especially crucial to estimate this cost when a change is intended to reduce manufacturing or field service costs. The design and development costs must be weighed along with the product, manufacturing, or field-related cost savings.

OPERATIONS AND FIELD SERVICE COSTS

Generally, the most significant (and most ignored) of all change costs is the manufacturing and field-support-related costs. These costs are not necessarily apparent to the engineer making the change.

Every impacted function may have associated costs. The supplier, purchasing, quality assurance, manufacturing engineering, production, materials, publications, etc., may all have one-time costs.

The field change labor, kit, repair, and retrofit-related cost must all be considered. Tools, fixtures, software, process/routing, test equipment, etc., may be impacted.

PART COSTS

Most engineers have a rough feeling for the parts and material cost of a change. They may not have a good idea as to what those cost will be under quantity purchases and actual production conditions however.

Although some companies make the design engineer responsible for knowing all the related manufacturing, materials, parts, and field-related costs, this author believes that to be an unrealistic expectation. Cost estimating will probably "take a back seat" to design work, or not get done in a quantitative manner.

A rail car company had the engineer estimate, the "change implementation cost" and enters it on their change order. We did an after-the-fact analysis on a sampling of changes and found costs to be underestimated or overestimated by a significant magnitude. Further analysis with a payback period established, indicated that at least 16% of their changes should not have been done.

COST POLICY

Many issues arise when discussing change costs. Should we cost all or only some changes? What costs should be included? Should costs that are normally part of overhead burden be included? Who will calculate the costs? Who will furnish labor and overhead rates? What build schedule should be used to annualize product unit cost?

The fact that there are so many perplexing questions undoubtedly deters the estimation of costs. However, it is imperative that all the associated questions be answered and it need not be difficult.

Policy: A standard is required to determine the company practices regarding change cost estimating. A form is needed to assure all the elements are considered.

Avoid creeping elegance. Failing to estimate costs is probably the single most significant CM-related reason for erosion of profit margins.

Policy: Cost estimate those requests which:
- **Are requested to "reduce time," for "ease of," or to "reduce cost" if any doubt about the payback exists.**
- **Are said to "improve" products.**
- **There are two methods of fixing a problem and we need to know the cost of each.**

Sometimes the requestor will have a proposed fix. Sometimes the engineer will need to roughly propose a fix before the cost can be estimated.

Some companies charge-back the cost to the requester or the party that saved the labor and material. This practice is typically an attempt to discourage changes. The result in this writer's experience is debate and finger-pointing. Better to screen requests with a payback analysis.

RESPONSIBILITY FOR ESTIMATING COST

The effect of the change on the supplier, tooling, fixtures, test equipment, manufacturing or service labor, etc., are costs that one should *not* expect the design engineer to estimate. You could call in accounting or industrial engineering and ask them to estimate the change cost. However, none of these folks are likely to have the "sense of urgency" that is needed. There is a better way.

Better, in this analyst's opinion to have those affected estimate their own cost and have CM gather their data, apply the appropriate overhead rates, and complete the payback analysis.

Policy: The executive champion must assure that CM has the resources to gather labor and other costs and to calculate the change payback.

The proper overhead burden rate for an activity-based cost (**ABC**) should be furnished by accounting. These are unique rates because some of the costs that need to be included (such as design time) are probably in current overhead rates. Each department should furnish their own labor and material information.

COST WORK FLOW

The work flow for cost estimating—assuming that a fix is known—is shown in Figure 11.2.

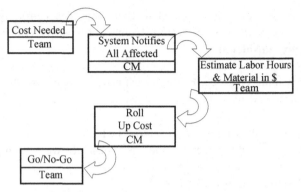

FIGURE 11.2 Work flow for estimating cost.

This process places the estimating responsibility with the functions affected while making CM responsible for "rolling-up" the estimates and calculating the payback.

While the flow diagram shows the team making the Go-No-Go decision, the final responsibility must rest with the cognizant engineer, while others on the team can take exception.

COST ESTIMATING STANDARDS

The system documentation would include the following standards:

- **Change cost policy**—specifies the kinds of changes to have cost calculated, estimated, and the company payback period.
- **Cost calculating form**—activity-based cost and payback form with instructions (see the *EDC Handbook*).
- **Cost form instruction**—box by box form instruction or cursor pop up instruction.
- **Cost work flow**—delineates the procedure via work flow diagram whether online or manual.

SUMMARY

Many mature companies make far too many changes to "reduce costs, save time, save material" when, in fact, they do not know if that will happen. Thus:

- **If you think estimating the change cost payback is expensive, try not estimating the payback.**
- Estimating payback doesn't need to be done on all changes, only those changes with doubtful payback.
- Set a length of time wherein you expect payback to occur.

- Develop the necessary standards.
- Accounting needs to furnish labor and special overhead rates for ABC cost estimating.
- The change team and CM need to make the estimating routine fast—and it need not be expensive.

Chapter Twelve

Change Management

Change management/control is often thought to be the beginning, middle, and end of EDC/CM/PLM. Of course, it is not the whole subject, but it is the single most important process in the entire product life cycle.

MANAGING CHANGES

A slow, error-prone or ineffective change process delays release (engineers hesitate to release in lead time if the change process is painful), adds costs, confusion, pain, and will likely give away a competitive edge.

Executives have sometimes pounded the table and declared "We have too many changes!" People may respond by putting more than one change into one change order, actually making the process more complicated, slower, and more error prone.

The mere thought of "control" strikes fear in the hearts of we engineers. Managers, service, and manufacturing people are constantly complaining that there are too many changes, except for the ones they request. We therefore need to seek minimum control and maximum screening of changes.

One of our greatest challenges is to find logical ways to sort out unwise changes, a subject previously covered. We also need to learn to do it right the first time! As we also discussed, it is wise to measure the number of requests and changes.

The changes per new designed document should go down over time if the release process improves. This is the most stated reason for improving the release process first.

The principles in the previous chapters must be applied if real reductions and reduction in the number of changes are to be attained.

Frequently folks know that the change process is long, painful, costly, divisive, and somewhat insane but no one seems to be responsible for the process. Because there are many functions involved, it is frequently stated, "Everyone knows how screwed up that process is!"

Policy: A single function, CM, should be given responsibility, authority, and resources to own and improve the change process with or without a small team to assist.

What a concept—process ownership! CM must be manned and budgeted to attain real improvement.

Configuration Management for Senior Managers. http://dx.doi.org/10.1016/B978-0-12-802382-2.00012-X

FAST CHANGE

Slow change control processes are the most frequent and insidious problems in the EDC/CM discipline. As already stated they discourage engineers from releasing documents in lead time. The slow change process also does the following:

1. Delays the incorporation of real cost reductions.
2. Delays the change that fixes a customer problem.
3. Delays the availability of a new feature or option.
4. Delays the fix for a field service/customer problem.
5. Increases the bone piles of down-level material.
6. Increases the rework or scrap.
7. Increases the work-around or line down time.
8. Increases purchased item cost because the suppliers have similar issues.
9. Increases the units that will require retrofit in the field (said to be 10 times more expensive than if changed in production).

Therefore:

Policy: The speed with which changes are processed shall be considered critical to profitability.

We need to develop *one* fast way to make necessary design changes. All changes that are worthy must happen quickly. If there are any changes that don't have to happen quickly, they probably shouldn't be done. This doesn't mean that they will all be made effective tomorrow, a different subject coming up shortly.

GET-AROUNDS

When the change process is slow, one or more fast ways around it are bound to be created, usually followed by the formal process. Thus, the change is documented twice, sometimes with a different fix in the formal change. Also the fast change configuration may never be reflected in the documents. The quick change processes simply take the pressure off the formal process. Doing it fast, followed by the formal change, is doing it twice! Therefore our efforts should be pointed at making the formal process fast.

Redline(s) on the floor are another frequent method of making a fast change. While the formal slow process creeps along, a "redline change" is done on the production floor with one or two signatures, sometimes even without any design engineer signing—go figure! *When truly necessary* to avoid a line down situation, why not make two redlines, have the author and acceptor sign both, get a change order number on the redlines, and require the now formal change to be in CM within 24 h? Do it once, fast, by the formal process. See the *EDC Handbook* for details.

There is no magic wand—most of the solutions discussed in this chapter and other parts of this book are needed to make the change process fast.

CHANGE TEAM

The change team should be made up of a technical representative from each potentially affected function. If face-to-face meetings are held, they must be frequent and well chaired. CM should probably chair the meetings. The "meetings" can be online only if our speed and quality guidelines are being met.

The team should consist of knowledgeable technical folks, not management people. As director of manufacturing engineering at a computer peripheral company this analyst was told he must join other directors on the change board. It wasn't likely that any director could be knowledgeable about all products and all issues so the directors were in affect "middlemen" whenever issues arose. The changes were delayed accordingly. Better to get the correct engineers in the process.

The time to bring a potential change to the team is right after engineering has taken ownership (request approved or manager directed). Before the design engineer has put fingers to keyboard the team should be made aware of the problem and discuss potential fixes.

The more technical functions involved in the problem and potential fix(s) the better. Not that all involved will be signers, but often they will have critical comment to guide the engineer.

Some change processes allow for a revised set of design documents to be presented to the team/board, probably already signed by engineering and rev updated. This creates an atmosphere (throw it over the wall) that says; "here, do it, we aren't interested in your thoughts or suggestions." With this method, the cognizant engineer may also think of any comments as "annoying second guessing." This is the number one reason for contentious team or board meetings. Swearing, finger-pointing, and even physical "finger punches" have been witnessed by this analyst.

There is a better way: meaningful teams discussing issues and then redlines for specific review and final documentation of the change.

REDLINE MARKUPS FOR CHANGE

There are three methods for documenting a change:

1. From–To or Is–Was, etc.
2. Marked up/redlined latest rev document(s).
3. Revised documents wherein the revised drawing rev block contains a description of the change.

The third method is seldom used (and wisely so) because it means that the change has already been made and is being "thrown over the wall" to the rest of the company. This method also means that every person who needs to understand the change (to cost-price it, to change manufacturing procedures, to change manuals, etc.) will be required to carefully compare the old rev to the new, which is waste of time and error prone.

The "From–To" technique is okay for very simplistic changes. This analyst thinks that it should be limited to about two or three lines of description on the change form. If the description is longer, then the redline technique should be used.

Neatly marked up documents—hand lettered in red—should normally be used because they:

- are easier to identify exactly what is changing;
- are considerably less error prone;
- avoid the throw-it-over-the-wall syndrome.

Most CAD systems have a redline feature. If that capability is not owned, it should be purchased post haste. Virtually all word programs have ways to exactly convey the change, showing both old and new.

The next step in the change process (team already reviewed the problem at least once) is for the engineer to bring latest rev marked up drawings and specs—either neat hand markup or CAD redline—to the team meeting.

> **Policy: Redlined latest rev documents will be the normal method used to define a change. They will be presented to the change team at the earliest opportunity.**

The double-spaced parts list redlined and the word documents (specs) clearly showing the old and new wording would be included as necessary.

The affected documents may be presented singly or in total, thus allowing for the more difficult issues to be discussed as soon as possible.

The engineer will look at the markup as a draft of the change rather than a finished version. The engineer will thus be more open to suggested alternatives, variations, or simply a tolerance change, from the team.

If supplier quotes or involvement is needed, redlines are the best method—hands down. If quotes or revised POs are needed after the change is approved and technically released, redlines and the change order should be sent to the supplier. This is much better than causing the suppliers to use a light table to attempt to find out what changed from rev J to K.

If customer review or approval is required, the team should discuss whether or not this is the time for doing that. The review or approval might be done and the change held or it might be decided to proceed with the risk. The engineering management might well use the *request* team for determination of the risk on a case-by-case or customer-by-customer basis.

ONE–ONE–ONE–ONE RULE

A few companies require one change order per document affected, thus they have multiple orders cross-referenced for some changes.

Many companies allow several problems to be fixed with one change order. This makes for more errors, a very confusing order, and makes meaningful metrics impossible. Each part of the change must be made effective independently because changes to hardware are not naturally batched. Product changes have different natural points of effectivity.

Volume change measurements require a method of counting changes that is consistent. The best rule therefore is:

- One problem
- One fix
- One change order
- One set of documents revised

The result is to focus the change order on a single problem, obtain a quality fix, process that fix quickly, and make it effective as circumstances require. This also makes change volume measurements most meaningful.

Changes to software are an exception to this rule because they *are* naturally batched. There are other exceptions to this rule which are covered in the *EDC Handbook.*

URGENCY/CLASS/TYPE

Some companies spend many hours debating the urgency of the change, usually because they haven't screened out the chaff. Thus, if you spend more than 10 man-minutes on determining the urgency of the change, go back to your request process and improve it.

Folks involved with DOD contracting or subcontracting are fixated on the Military Class I, Class II, and records changes. When required by contract, do this for your customer, but recognize that it is somewhat meaningless to your change process. Class I is essentially noninterchangeable (although never properly defined in Military Specs) *and* Class I includes price increases. Records change is a document only change (which is meaningful) and Class II is all the rest, which should be interchangeable changes, except for the lack of clarity regarding the definition of interchangeable/noninterchangeable.

Sometimes folks try to classify changes based upon the disposition of the old parts or other criteria, usually resulting in somewhat meaningless results.

The most meaningful method of classifying changes according to this analyst is:

Class:

- Document only change (no affect on the item: Mil records change)
- Interchangeable
- Noninterchangeable

This classification allows for simplistic processing of document only changes and wise decisions for part number changing and change tracking.

Type:

- To meet product spec (noninterchangeable by definition)
- Improvements over and above product spec (interchangeable)
- Cost savings

This allows for an urgency to be applied to the incorporation of changes to meet spec and a wise treatment of material disposition and retrofit.

The change form should contain check boxes for both class and type. For example; if a change is checked to be interchangeable but is also said to be required to meet spec, the CM technician can immediately raise the flag for apparent conflict.

Remember that part number change on noninterchangeable change is a separate consideration based upon whether or not units have been shipped.

SOFTWARE CHANGES

As covered previously, the CM organization does need to control any transmission to a customer, plus all code releases, both initial and changes.

The software "releases" are of two kinds from a product CM viewpoint. The initial release needs to be done through product CM just as any other part of the product (see Release chapter). Each subsequent code "release" should be done through product CM documenting the embedded software changes by a change order with:

- Redlined spec including:
    - Changed part number, if tabbed, 02, 03, etc.
    - New version number of the software code
    - Used-on product numbers and PNs
    - Build environment, tools, settings, and other pertinent data to allow retrieval of the code and regenerating the media

- Two copies of the revised media: one for production and one for the CM file, marked with:
    - Part number (same as above)
    - Name
    - Version number

CM will assign the next rev to the revised spec. The change order form can be the same as used for hardware changes or unique. It should contain a list of all requests that have been included in the release. Changes to software are correctly and naturally "batched." This makes the new release noninterchangeable by definition, thus the PN changes to 02, etc.

Ideally the change order should also list the requests that are approved but *not included* in this change/release. If CM is maintaining the software request log, this isn't necessary.

MANUFACTURING, QUALITY, AND SERVICE DOCUMENTS

Many companies bundle the fabrication, assembly, inspection, test process changes, service manual changes, etc., in with the design documentation change order. This practice causes the change to wait on those redlined documents unnecessarily and unwisely.

Sometimes ISO or, FDA or other certifiers/agencies imply that nondesign documents must be in the design change order, but this is simply not true.

While waiting for nondesign documents, all the negatives mentioned earlier occur. While waiting, another change to one of the documents involved is often necessary and the change order is revised to make that additional change. The ME, IE, TE, QE, etc., now all have to re-redline to the latest modification to the package. Round and round we go with each blaming the other for the delay. This practice is a touch of insanity.

The design change needs to be completed first and the other documents completed as a second step in the process. This will speed up the process and eliminate considerable finger pointing; as seen in Figure 12.1 the support documentation changes should be a second step.

This does not mean that the nondesign folks cannot do some of their work in parallel with a design change. If the team is functioning well they can *begin* their efforts as the design change is being developed. But they cannot, in fact, *complete* their efforts until the design change has been technically released.

The other function's technical documents often need to change for reasons other than a design change. These should be done without use of the engineering change order.

Using this method will, of course, require a way of assuring that the other documents are updated according to the design change. Quality assurance should assure that this is happening and CM/exec champ should see that they have proper methods in place.

All the documentation which ships with the product (or is online about that product) must match the design of the physical product. This would seem to this author a natural part of the QA responsibilities. If CM is tasked with this responsibility, it needs to be manned accordingly.

Sometimes even quality assurance identification of a "miss-match" doesnt work. One computer manufacturing company had a deviation hung by QA on every product shipped stating that the publications didn't match the product. This had been going on for years. A little investigation revealed that the publications people were across the city in another plant, which made different

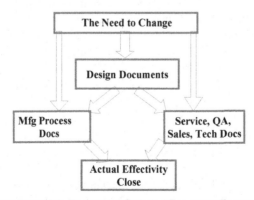

FIGURE 12.1 Sequencing Technical Document Changes.

products. Their product publications naturally did match the changes. The engineering services director made the chief engineer and general manager aware of the situation. The solution was to transfer the people required and move their work place to the proper plant, a painful but necessary action.

IMPACTS OF CHANGE

Some say it is necessary for the cognizant engineer to be aware of the exact departments affected and all impacts of the change on them. This writer believes that this might be practical in very small operations or if the engineers have gone through extensive company training.

Caterpillar Tractor Company put this engineer through 9 months of on the job training by serving 1–5 weeks in all functions related to my eventual position. If one didn't know the impact of the change, you certainly knew who to ask. Few companies put new engineers through such exhaustive training, however.

The best practice for most companies is to have an impacts list on the change form with an instruction that asks the engineer to check-estimate who is affected. Then all involved, after technical release of the change, must review that initial estimate and take responsibility for the correctness of the impacts checks, 1 work day allowed.

Policy: If impacted functions do not respond within 1 work day, they, by policy, will have taken ownership of engineering's judgment.

It is then the impacted function's yes or no decision. This must be the policy if a fast change process is to be attained.

TECHNICAL SIGNATURES ON CHANGES

A seminar attendee reported that they had every function that *might* be affected by *any* change sign the change order. They had to obtain 14 signatures on every change order. At one company this analyst witnessed 11 signatures on every change document. This supposedly "fail-safe" method is, in reality, a touch of insanity.

The number of signatures based upon the author's 58-company survey varied from 1 to 12 with 5.4 the average. The company with a single signature was an "engineering-driven" railcar manufacturing company. They were experiencing many assembly, fabrication, and supplier issues as a result. The companies with more than three or four signers were, no doubt, suffering with a very slow process.

Few companies have anyone signing redlines, where all the technical issues reside. One primary internal customer (usually operations) should be signing redlines to assure manufacturability, testability, etc. Signatures on the change order document should be administrative signatures plus that of the Cognizant design engineer.

As previously discussed, the one author–one acceptor method is most desirable. Each redline document should be signed by the engineer author and primary acceptor.

Have the responsible engineer obtain that primary customer signature (rather than CM) to assure engineer-to-engineer discussion without a third party "in the middle."

Policy: The design engineer should obtain the required signature(s) on redlines in a team meeting/online prior to giving the change to CM as technically complete.

If the change is done by Was–Is/From–To technique, the signatures will have to be on the change form.

The cognizant engineer should sign both redlines and the change form because the change form will contain where used, products affected, etc.

The Just-in-time manufacturing principal should be used wherein:

Policy: Anyone affected by a change order has the authority to stop the process.

E-mail the CMs and they will resolve the issue. This is the best method for limiting signatures on changes. It also encourages fast review by all involved.

CHANGE PACKAGE

The technically completed change *package* must contain:

- Redlines of those pages of design documents changing and all associated pages of related spec documents changing.
- Design documents defining any new parts required for the change.
- Rework instructions—if not simply "rework to print."
- The change form technically completed. The technically completed change form must contain:
  - The associated request number if applicable
  - Request accepted date or the date the engineer was tasked with initiating the change
  - Justification for the change (better word than "reason")
  - Class and type
  - Any customer-driven effective dates or serial number requirements known
  - If field failures, the number of units that have failed, field population, customers involved, safety issues if any, etc.
  - If customer/agency approval or notification is required, date done or plan to proceed with risk outlined
  - Check the impacts list with best estimate of those thought to be affected. The checks corrected by those affected if necessary
  - From–To description of the changes including drawing zone, if brief
  - Lab testing: SN modeled/tested and test report number
  - Old PN & Rev, new PN, noun name, suggested old part disposition
  - Indicate if the old PN can be physically reworked into the new PN
  - List of applications/products affected (used-on)
  - Retrofit plan if applicable (developed with the field engineer)
  - Cognizant engineer's signature and date

Note that if retrofit is required the decision is made during the design team discussions. Failure to make field retrofit decisions during the change process leads to an abdication of the decision—usually to the service function—often excluding the engineering function and the change team. The result may be costly retrofit of changes that may not need to be retrofit.

For detailed form and form instruction see the *EDC Handbook*. CM must ensure the completeness of the change order at technical release.

TECHNICAL RELEASE: POINT OF NO RETURN

The point at which the change is technically complete is often unclear. The fix can be modified at any time. This process allows the engineer to launch untested or incomplete changes and start–stop–correct (perhaps over and over) all the implementation activities. There are several questions. When should CM update the systems? When can the IE safely revise the assembly process? When can the test engineer revise the test process?

The result without a clear tech release point is that folks wait until the change has been incorporated into the systems and the master documents before they start implementation. This yields the slowest possible implementation process.

Avoid this common change process fault. It must be very clear as to when the engineers responsibility ends and the implementation can begin. At this event, CM must do an immediate inspection of the form and redlines to assure that the standards have been met. They should do this in an average of 1 h or less.

If not met, the change should be immediately returned to the engineer. If met, clock this occasion as tech release.

Policy: Once a change has been accepted by CM for tech release, any issues/ problems with the change must be corrected by a different change order.

This analyst sometimes refers to this as "the drop dead point": if the engineer contacts CM with a correction to the change order, he or she is told to "drop dead," write another change to do the correction. This is a necessary and critical practice for fast and painless change processing.

Of course there will always be exceptions but the normal process should require a new change order and exceptions must be rare. The reason is very simple: engineers and CM must learn to do it right the first time, not to practice the age old "launch something in a change and then go back to finish testing, customer satisfaction, etc."

Some issues needing correction will be of CM's doing and they should write the change order to correct the problem. Others will be design issues that the engineer should correct with a new change order. You will see shortly how this will enter into the change quality measurements.

CM must notify everyone involved that the change has been technically released. This will mean that they can start their individual activities to implement the change with very little risk of wasting their efforts. Only administrative issues remain.

All who need to complete their implementation efforts need to be confident that a very low risk point in the process has been reached. The "flag" for this event completion is the assignment by CM of the next rev level to the affected documents.

REV LEVEL

Rev level assignment after the release to pilot production must be in the hands of CM. CM must assign the next rev only upon the change passing the tech release check.

> **Policy reminder: If CM doesn't have control of revs, you do not have control.**

Many companies fall into the *Rev Rolling Trap*. Anytime a parts list is affected by even one component rev change, the parts list of every using assembly must be revised to increase their rev simply because the list has a column for rev level. Then all assemblies using those assemblies must have their rev Level increased, on ad nauseam to the very top (product) levels.

> **Policy: Our policy shall prohibit Rev Rolling on assemblies wherein there are only component revs changing.**

Methods to avoid this totally unnecessary activity are:

1. to develop a policy that simply outlaws Rev Rolling or
2. to develop the engineer friendly parts list (Figure 9.3)

Sometimes folks mistakenly believe that rolling revs results in change tracking, but it doesn't, unless a complete bill of material for each unit is identified by serial number and kept forever. Even then it is a very impractical way to trace a change to say the least.

If an engineering friendly parts list is programmed—double spaced for markup and without a column for component rev level—then the policy outlawing the practice won't be needed.

UPDATING THE MASTER DOCUMENTS

Update of the master documents to incorporate the change is a critical function that needs to happen immediately after technical release of the change.

The typical "less than best process" calls for the following:

1. The approved change is returned to the designers or drafters (not in CM) for incorporation into the master.
2. The change incorporation is somehow, sometime fit into the new design workload, usually after CM badgers the designer relentlessly.

3. The updated masters are then returned to the engineer to sign the rev block.
4. Sometimes the designers add new changes or modify the change when they are updating the masters. Sometimes engineers make changes or additions when they are called on to sign the rev block of the updated master.
5. Often a rule is made that the ERP won't be updated until the documents are revised, signed, and released. This rule then creates an atmosphere that discourages any parallel implementation activity to shorten the process and implementation time.

Certainly, this is a less-than-best process. This analyst would have CM do the update of both the BOM file(s) and the CAD masters form the precise redlines.

Policy: The designers/drafters who incorporate the changes into the master documents should be part of the CM function.

If the incorporation drafters are not part of the CM function, the change process speed will be significantly affected. If the approved change goes back to the engineer/designer/draftsman for incorporation (a person/function also responsible for new designs) bad things happen;

- The change will take low priority in their work. It is simply human nature to want to work on new design rather than on changes. If the design/draftsmen who will incorporate the changes are placed in CM, it will be their only and top priority.
- The longer it sits there, the more the temptation to alter the change or "piggyback another fix" into the change. This is a major contributor to change complexity, slow process, and frustrations with the process.

Policy reminder: One problem, one fix, one change order, one set of drawings revised.

When the change is technically approved (technical release), the CM organization will assign the next rev and incorporate the change into the masters, without delay and exactly according to the approved markups.

This presupposes that the portion of designer/draftsmen who are currently doing the work will be transferred to CM. If the markups are done in CAD, the incorporation of the change into the master document usually requires only a few keystrokes.

As previously discussed, CM will also incorporate the design data changes into the ERP/PLM systems. Thus we can eliminate the tendency for BOM update to wait on design document update (or vise versa in a few companies). Both must happen with an average of five work days. Many CM functions have proven that they can do that, some even faster.

One electromechanical device client reported that they had implemented a new system and had achieved a 3-day average CM time.

Part of the CM activity must be to find out the initial plan for the effective date in each product affected by the change.

QUEUING CHANGES TO THE MASTER

There is a practice sometimes used which avoids changing the master document until several changes accumulate. Mil standards, ISO, and most agencies allow for this to happen. The drafting room manual allows this and names the practice "Advanced Document Change Notice" (ADCN), up to five changes before master update. There has never been and there is not now anything "advanced" about this practice. It is simply "retarded."

> **Policy: If your standards allow the ADCN type practice, eliminate it immediately: one problem, one fix, one change order, one set of drawings revised, immediately.**

The very idea that all the users of a document could somehow accurately incorporate changes into their copy of that document before they can use it is simply absurd. See the *EDC Handbook* for a complete discussion of this insane practice.

EFFECTIVITY

CM folks have coined a word not found in the dictionary: effectivity. It is a unique term for a complex subject: when, and/or in what serial number, and/or at what point in the process shall the change be made effective. This prompts other questions:

- Shall we make the change earlier in time and thus have fewer units to retrofit?
- When can we get the new material or parts required?
- Should we pay premiums to make the change earlier than practical?
- Shall we rework or get revised parts? How much rework?
- Will the effective date be adequate or will we need to know the exact units affected.
- Who or what is the pacing item? Are the fabrication, assembly, test, or service procedures pacing? Tool design/modification pacing? How about the publications? Are the materials or part purchases pacing?
- What are the customer wishes and requirements?
- Who will sort this out?

Failure to know at least the approximate units affected/not affected (via the date effective) leaves the service and design engineers with a quandary when later dealing with a failure problem. Given a failed unit(s), did a change fix the problem or could a change have caused the problem?

The usual and best practice is to have production control coordinate this task. They are already involved in all scheduling and materials and parts issues *which pace most changes*, but other functions sometimes pace the effective date of a change.

The technically released change order is in the hands of all potentially affected. They have taken ownership of the impact on them, yes or no. If yes, they will be allowed one more day to contact the PC to tell them what they have to do and when they can be done. This may be done via online work flow or e-mail.

PC can thus tell what the pacing item is (including material availability) and set a date for that item to be completed. They would then notify CM as to the planned effective date. CM would enter that date into the change order and into the ERP for PL/BOM changes.

If the online process allows, PC might well enter the date into the BOM and change order and update it as necessary.

The first effectivity plan should be on the change order as well as subsequent adjustments of the date and the actual date.

EFFECTIVITY VOLATILITY

As we well know "the best laid plans of mice and men aft gang agley." Especially volatile are the parts and materials availability. PC must follow every open change to revise the plan when necessary. Every impacted function must notify PC if the plan changes. PC must, in turn, notify CM (or the work flow system) of any pacing item change. CM (or PC) will update the change order for all to see and input the new replanned date to the ERP for driving materials requirements.

When the change is actually implemented, that date must appear on the change order for posterity. Not only does the ERP system need to be kept up to date with the latest plan but the change order will show the plan, replan, and actual effective date. Why? Because proper materials planning requires it, all involved need to be able to access the latest plan in order to plan their update work and it must be available for troubleshooting. The BOM should also have ability to store and display the historical information for troubleshooting and liability purposes.

Realize, however, that such "actual" dates do not directly relate to specific units affected. For a variety of reasons, the exact units affected are unknown. The best we can do with the actual date is to approximate the units affected. This is satisfactory for interchangeable changes for most agencies and commercial practices. However, best commercial practices require noninterchangeable changes to the exact unit. NASA and some others require tracing of *all* changes to the exact unit.

TRACING CHANGES

We have now captured the actual effective date (and thus the approximate units) for all changes to satisfy most engineering troubleshooting needs. Good commercial practices as well as some customer requirements call for knowing the actual units affected in certain cases.

Certainly any change for a safety issue or one to be retrofit would prompt us to trace to the exact unit. This analyst believes that all noninterchangeable changes should be exactly traced. Customers and agencies have unique rules for tracking or "status accounting" some or all changes. See the *EDC Handbook* for details.

> **Policy: Trace all changes that are noninterchangeable and those required by the customer to the exact SN/Batch/Lot/Order as applicable.**

This policy is easier said than done, costly too. The expenses involved are the reason that this writer would not track interchangeable changes unless the customer or agency requires it.

Some commercial companies have chosen to trace all changes to the exact unit, probably because they don't clearly understand interchangeability.

WHO AND HOW TO TRACE

Just as it is natural for production control to set effectivity, so it is natural to look to them to be responsible for tracing changes. The CM manager should work with the PC manager to establish an acceptable method.

The most common practice is a unit traveler wherein the designated changes, when incorporated are logged in the traveler. Then the individual data is input to a record available for searches. Other methods are discussed in the, you guessed it, *EDC Handbook*.

The results should be in a database report which is available for all who need to know.

OLD DESIGN PARTS

Failure to properly disposition old design parts during the design change process contributes to the "bone pile" of down-level material in operations. Often thousands of dollars of material exists in limbo which should have been used, reworked, returned to the supplier (if done on a timely basis), or scrapped. Is the supplier more likely to give some return credit today of some weeks or months later? Thus the engineer on the change order should suggest a "disposition" of the old design parts line by line as follows:

- Scrap
- Return to supplier
- Use as is
- Reworkable

Notice that the engineer should not be expected have the final word on whether or not the items should be reworked nor returned to the supplier. These are properly an operations decision. They have schedule commitments to consider. They may determine it most cost effective to rework or perhaps to obtain new design parts and return or scrap the old design parts.

Return to supplier is a disposition that has all but disappeared in today's manufacturing. It is about time to bring it back.

Policy: Operations folks will determine whether to rework or not and whether to return to the supplier or not.

Their decision may be quantity sensitive and should be noted on the change order, PC via e-mail to CM or into the online system.

CLOSING THE CHANGE

Most companies close the change after the BOM and the master documents have been updated—this it too early. We need to know that the change was incorporated and when or in what units as applicable, for the reasons previously stated.

Another reason for tracking the closing of every change is for the very unusual circumstance that this writer encountered when consulting for an electromechanical device manufacturer. When walking through the change process, a count was made of the changes in process to compare with the CM database. Discrepancy of a dozen changes was found. After pondering over that discrepancy for a while, we went to operations where the changes were sent for incorporation. At the desk of the responsible person we found two stacks of change orders. We found that one pile was not yet incorporated and the other pile found was; "we won't be doing those because we don't think they are necessary!"

Further discussion revealed that recent changes to the process caused the operations folks to feel shutout of the process. When they were confronted with changes "thrown over the wall" they simply made their own decision as to the worthiness of some changes. The CM manager quickly got them involved in the process.

In any organization, there is probably room for some change(s) to be forgotten in the implementation process, so tracing to implementation will negate that possibility.

OBSOLETE

When is an item to be considered obsolete? First realize that operations may have a different definition of obsolete material than engineering . Terms must be carefully defined. Engineering usually defines obsolete as "not to be used in new designs."

Engineering obsolescence should occur by a change order using the normal change order process. The cognizant engineer will normally note in the document rev description block and/or in a prominent location on the drawing/spec the item number that should now be used, such as "Obsolete for new designs – use PN XXXXXX."

A used-on report should be attached to the change order to assure replacement of the obsolete item in all using assemblies. This should prompt all the normal implementation activities including possible return to supplier.

CM will change the rev of the document, in our case we chose to use OBS in the rev field.

If a *product* is to be made obsolete, it should be done by change order. Engineering or CM should check the used-on for *every item in that product*. Any item unique to that product should also be considered for obsolesce and items in stock disposed of.

CHANGE STANDARDS

Documenting the standards not only provides a training tool but also a basis for process improvement. The change standards that most companies should have in their CM standards manual are:

Change control policy: Defines the policy and practices required.
Change process flow diagram (procedure): Describes significant process events, the sequencing of those events, and the responsible department for each event.
Change process teams: Best practices, membership, and responsibilities of the team.
Change form: Hard copy and/or online.
Form instruction: Requirements for each form box. Pop-ups online.
Part number and rev level: When and who will change part numbers and rev levels.
Change class and type: Defines the acceptable and required change categories.
Markup of design documents: Specifies the required methods for precisely defining the differences between old and new document rev levels.
Effective point/date: Defines the points in the manufacturing process and the date the change will be made effective.
Effectivity management: Defines who and how effective date (or other) planning and actual effective date or SN management will be done.
Disposition of old parts: Specifies who will be responsible for and how old design parts will be disposed of and the acceptable categories of disposition.

Again, apply the principle of one subject, one standard, very few pages in each.

CHANGE WORK FLOW

Design of the change work flow needs to be carefully considered. The clarity of engineering, CM, and operations responsibilities is critical. Yes, there are many activities in all phases of the process but the basic responsibility for the major phases of the process, and all the events therein, need to be very clear. See the *EDC Handbook* or *CM Metrics* for a detailed work flow diagram.

This analyst's flow is;

Phase	Basic Responsibility	Number of Events
Design phase	Engineering	12
CM phase	CM	13
Implementation phase	Operations	10

An abbreviated change block diagram, showing the major points in the process to clock, and some of the events included, looks like Figure 12.2.

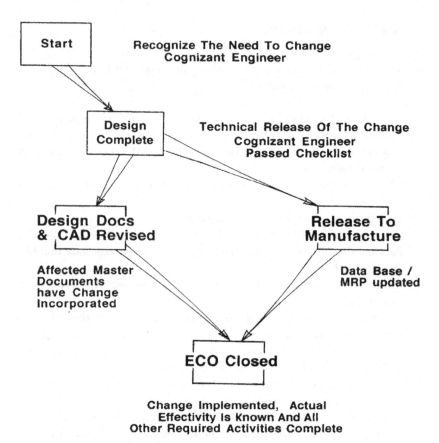

FIGURE 12.2 Block diagram of major change control events to measure.

Notice that neither the update of master documents nor the update of the ERP/ PLM systems waits for the other to be done, they are done in parallel. Since they will both be done quickly, the need for either waiting on the other is eliminated.

Block diagrams like this are nice for dog and pony shows and for this book, but detailed event work flow *showing the function responsible for each event* are mandatory for process understanding, measurement, and improvement.

CHANGE METRICS

The *CM Metrics* book contains many examples of useful change process metrics. We will cover here, only the most useful and enlightening measurements.

The volume and use of deviations is critical to the change process for reasons previously discussed. The use of deviations for making a fast change should be eliminated by making the normal change process fast.

A metric (duplicated in Chapter 5) should be put in place to track deviations as in Figure 12.3.

If other methods for making fast changes exist, they too should be eliminated in favor of one normally fast change order process.

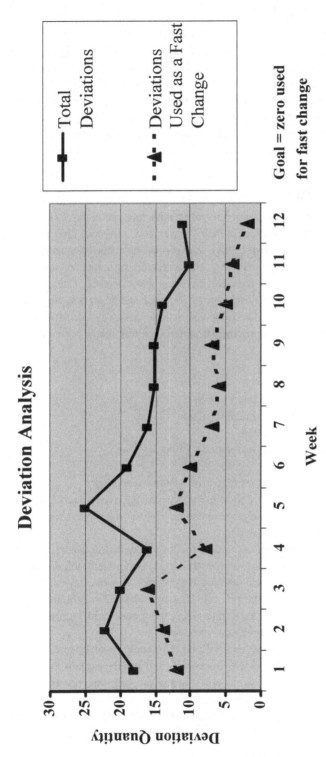

FIGURE 12.3 Eliminate deviations used for fast change.

The change process time by phase should be measured and reported to management and all involved. Figure 12.4 shows such a metric by month.

This metric should be prepared and distributed weekly if possible. The trend in this case is positive. This writer has no benchmarks for the design time (engineering phase) or implementation time (manufacturing phase). Continuous improvement should yield continuous decline in the parts and in the total time.

The most important measurement of all CM metrics is, in most companies, the change processing time. This process measurement may be best examined by study of a particular situation—thought to be all too common.

CASE STUDY

The Fortune 500 computer peripheral manufacturer (referenced earlier) knew they had a process time problem in their "money cow" division. They had few metrics except volume measurement, they were processing about 100 changes each month. They knew that the process was loaded with pain, suffering, and complaints. Emotions ran high whenever the change process was discussed. Change control board meetings were fraught with arguments, swearing and finger pointing. The new executive VP of the division decided to take action. This analyst became involved.

They started with time measurement of a few major points in the process. They measured:

- Start date
- Changed design input to CM
- Date master documents updated
- Date BOM updated
- Date the change was actually effective

They found that the average process time was about:
38 work days from start to giving the change to engineering services (CM)
40 work days from change to engineering services to master documents and BOM updated
41 work days from last update to close (actual effective date or SN known)
They also tracked the volume of changes completed each month.

This was the "biblical 40-40-40" process—40 work days and 40 nights for each major portion of the process. This was about 2 months for each phase or 6 months total. We decided to reengineer the process with emphasis on reducing the middle 40 as it seemed most excessive.

A short meeting was held with all the people involved to explain the problem, discuss why speed was important, and briefly explain the project. They put three key people to work with this analyst, full time, one from engineering services, one from materials, and one from operations.

Large 2 ft × 3 ft process time graphs were put on the wall in engineering services, in the cafeteria, outside the chief engineer's office, and outside the executive VP's office.

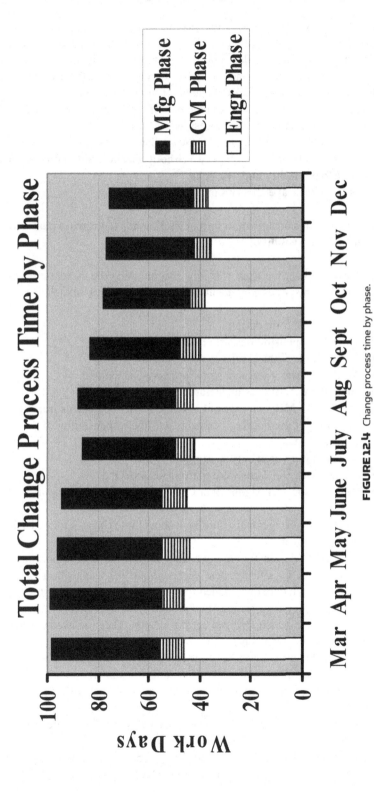

FIGURE 12.4 Change process time by phase.

A couple of weeks following the meetings, the process time decreased 2 or 3 days in every part of the process. The team had made no process changes as yet. We met and discussed this time reduction. Our conclusion was that when folks understood the importance of speed, they took it on themselves to help. Within another couple of weeks, a couple of more days came off the front and rear portions of the process. The middle part of the process came down another 6 days, again, no changes to the process as yet.

(A farm implement manufacturer saw similar but lesser results by adding to their EC form, in bold letters, **"Speed Without Sacrifice of Quality is Important to our Profitability and our Jobs."**)

Thus the industrial engineer's most basic rule:

Principle: Measurement, in and of itself, tends to improve performance if given high visibility.

In the meantime, the team was flow diagramming the current process and critiquing it. Many issues were identified including (but not limited to):

- Poor change form design.
- Unclear responsibilities with multiple signatures.
- No point of technical release was evident.
- Work flow was a series of steps, little done in parallel.
- Input to the MRP was in operations.
- Input to the MRP waited until after the master documents were updated.
- Updating of the master documents was the responsibility of the design groups.
- Redlines were used only sparingly.
- The existing standards were sparse and confusing.
- There was no training in either CM or other involved functions.

A sampling of changes and a data bank about those changes formed a further basis for improvement.

A team goal was set for improvement: reduce the middle portion of the process to 5 work days in the next year without increasing the front or rear throughput time. The goal was added to the graph. The die was cast. Cooperation was outstanding because the chief engineer and executive VP cochaired a steering committee which received regular reports. They removed obstacles as they appeared in the team's path.

Details of this project can be found in the *EDC Handbook*. The metric for this project is shown in Figure 12.5 with results by quarter.

As you can see, the team essentially met its goal in about than a year and a half. This performance was achieved with only two data processing systems, MRP and CAD. The CM process was what today would be referred to as a purely manual/hard copy system.

The engineering redesign time and volume and the operations implementation time and volume were also measured. The front end and rear end of the

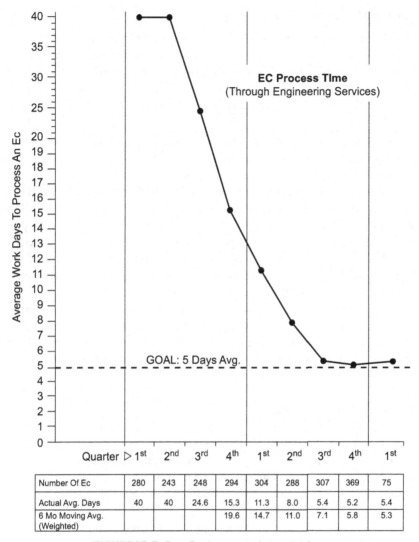

	Quarter ▷ 1st	2nd	3rd	4th	1st	2nd	3rd	4th	1st
Number Of Ec	280	243	248	294	304	288	307	369	75
Actual Avg. Days	40	40	24.6	15.3	11.3	8.0	5.4	5.2	5.4
6 Mo Moving Avg. (Weighted)				19.6	14.7	11.0	7.1	5.8	5.3

FIGURE 12.5   Case Study process time and volume.

process were also speeded-up by a few work days not withstanding the fact that some tasks were removed from the middle and put up front or out back.

The quantity work in process (WIP) was also measured (bottom chart) in these metrics since it is a good indicator or future throughput time.

In retrospect, some individual elements of the change process should also be time and volume measured in most companies:

- **Change incorporation into the master drawings and specs**
- **Change incorporation into the ERP/PLM systems**
- **Effectivity planning**

Without a doubt the change process needs to be measured for quality.

CHANGE QUALITY METRICS

There are many people and functions involved in all phases of the change process. People are prone to make 2–3% error (as learned somewhere in the writer's IE training). It follows then that several people involved in one phase of the process can produce a double-digit error level fairly quickly. Also, too many companies have a change process and culture which produce a negative effect;

> **Negative principle: We can't find time to do it right, but we can always find time to do it over.**

The best method of preventing errors is to have well-trained people. Standards must be available as a basis of training. The next step is to require each person in the process to be tasked to check the work of the person preceding them in the process and to make sure that the same person who made the error, corrects the error.

When this analyst was put in charge of the engineering services at a major computer company, he found that when errors were found the person who made the error was not informed about the error, let alone required to correct it—a touch of insanity! It should be obvious that the person who made an error should correct the error.

The best practice for measuring process quality this analyst developed with a client is a measurement of error corrections. This measurement includes revs to a change in process (if allowed) and a change to correct an earlier change.

Corrections should be divided into two parts:

1. **Design error corrections:**
   a. Fixes to the design fix after the markups have been signed and tech released (if allowed).
   b. Change orders to correct design errors in earlier change orders.
2. **Administrative and technical error corrections:**
   a. Fixes to the administrative aspects of a change document after tech release of the change.
   b. "Document only" corrections after incorporation of the change into the design documents or BOM.
   c. Errors found in the incorporation of the changes by CM to the design documents or BOM, whether corrected by revising the change in process or by another change order.
   d. *Exclude* replan of effectivity dates.

With those definitions in mind, the CM folks should merely count the number of errors found in a week or month and compare that to the number of changes made during the same time frame. Given that the corrections may have nothing to do with the changes made in the same time frame, the result is a quality control or **QC factor**.

> **Example:**
> Design process quality: 5 fixes this month ÷ 20 changes this month × 100 = %25 design QC factor.

Kept over time we can find trends. For an example of the design team metric see Figure 12.6.

A similar collection of data for the CM phase:

Example:

Admin process quality: 4 admin revs this month ÷ 20 changes this month × 100 = % 20 tech QC factor.

Over time this measurement might look like Figure 12.7; Graphed over time the result will show a trend; in the case of Figure 12.6 and 12.7 the trend is positive.

The client company where this author and the CM manger developed these metrics had results much like those graphed. The other (somewhat rare) feedback the author has received from other companies have reported similar results.

Imagine the time saved when 20% and 25% of the changes needing error correction is reduced to 3% or 5%—time to address worthy requests and process improvements. In both metrics, 1% or 2% should be an attainable goal.

It must be recognized that there is potential evil in these metrics. It is all too simplistic to interpret them as a measure of design engineers, design departments, the CM department, or the design group as a whole. There are far too many people and departments involved in the process to say that it measures one person or one department. The first measures the *change design team* and the

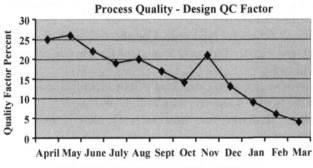

FIGURE 12.6 Change Design Team QC Factor.

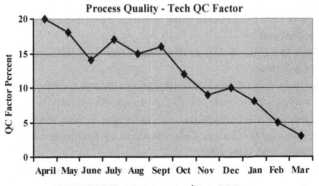

FIGURE 12.7 Administrative/Tech QC Factor.

second measures CM/PC/change incorporation/drafting portion of the change process. See *CM Metrics* for further explanation.

> **Policy: These change quality measurements and reports should be put in place immediately.**

> **Policy: The QC factor quality measurement must be about measuring the process or team, not individual people or individual departments.**

Of course, the graphs have to be given high management visibility.

SUMMARY

The change process is broad and deep, important and complex, but can be simplified, effective, and efficient if addressed in small bites and chewed well:

- **Get key measurements in place as the first step toward improvement: volume, time, WIP, and quality.**
- **Decide whether to reinvent or continually improve.**
- **Make sure that CM is chartered and manned to improve the process and to make it fast and effective.**
- **Develop a standard for each small bite.**
- **Put an effective change team in place that meets face to face until our benchmarks are met.**
- **Have all functions represented on the technical review team, allow very few signers but have a method for anyone affected to give a "stop order."**
- **Determine how those impacted will be identified and involved.**
- **Decide to depict changes with redline markups.**
- **Resolve that one fast process is all that is needed—eliminate get-arounds.**
- **Use the one–one–one–one rule.**
- **Use classes and types as suggested.**
- **Make the design change first and the other technical document changes a second step.**
- **Assure that the process work flow has a point of no return, technical release point.**
- **Move the folks doing design data input to ERP and PLM into the CM function.**
- **Move the folks doing master document change incorporation into CM.**
- **Get production control onboard to handle effectivity and tracking.**
- **Determine that firmware and software development are basically the same as mechanical, electrical, or hydraulic development from a product CM viewpoint.**
- **Toot your horn when significant progress is made and celebrate the same.**

Chapter Thirteen

Field Change Process

The field change process is actually an extension of the change process. However, many companies do not incorporate changes in shipped product. Thus this chapter is kept separate so that folks not interested can skip it.

Customer service is essentially the final customer satisfaction test. A product delivered prematurely to customers can be "saved" by outstanding field service.

As mentioned before, the determination of whether or not to affect field units should be done during the change process. The type of retrofit should also be stated in the change order. For those that do make some field changes let's explore the essentials of the field change order (FCO).

SAFETY RECALLS

When a safety issue is evident, immediate action to initiate and implement a fix should be taken. This should not be the only action taken immediately however.

- If any design or testing work needs to be done, give a heads-up to the customers or dealers outlining the problem and the action which should be taken by the customer.
- Notify all customers with a phone call and a letter.
- Put notice on the company Web site.
- Run ads on related sites, magazines, radio, television, and newspapers as might be prudently expected to reach your customers.
- Write a "recall" FCO as soon as possible with complete instructions for action to be taken.
- Keep careful track of every action taken—who, what, how, when, where, and how much.

Leave no stone unturned to avoid letting an elephant (litigation) into the room.

CHANGES TO RETROFIT OR NOT

Which changes should be considered for retrofit?

- **Policy: No interchangeable change will be retrofit.**

Configuration Management for Senior Managers. http://dx.doi.org/10.1016/B978-0-12-802382-2.00013-1

- **Policy: *Not all* noninterchangeable changes will be retrofit.**

One major computer company saved a million dollars a year (in 1965 dollars) by simply ruling out the retrofit of interchangeable changes. They had established a "retrofit to the latest rev" policy back when essentially all changes were to meet specs. They also went on to sort out noninterchangeable changes which didn't need to be retrofit and saved another boatload of greenbacks.

Design engineering, CM, and field services need to jointly examine the possibilities carefully. The classification of the field change should be stated in the change order since it is necessary to know the class of retrofit to calculate the change cost properly.

CLASSIFICATION OF FIELD CHANGE

The effect on the field units should be determined in the change process. The field service folks should cooperate with CM manager to determine the types for your company. Types of field effect are as follows:

- Recall
- Immediate
- On failure
- At regular maintenance
- Others

Of course there are other potential types that include sale of a kit for repair, customer use of online instructions, etc.

FIELD CHANGE ORDER

There are several unique features of a field change that need mentioning which are as follows:

- Disassembly instructions are normally required.
- Reassembly instructions are normally required.
- Testing instructions are often required.
- A kit part number should be included.
- The FCO document should be given a PN.
- The FCO should be assembled in the kit.
- Include the serial numbers to be affected in the instruction.

A PN for the instructions (FCO) allows the FCO document to be stocked, issued, and assembled in the kit. The effectivity serial numbers (or equivalent) allow the field service folks to zero in on the units that need the change.

Several other aspects are discussed in the *EDC Handbook*.

FIELD INSTRUCTION WRITING

It is critical for the instruction writer to understand that field service folks have an axiom: **When a problem is experienced in the customer's environment, the factory is always closed!** This writer found this out the hard way, as manager of Worldwide Repair Centers in a fortune 500 company. The customers or field engineers have their training, product publications, their kits, and their wits to maintain the product.

> **Policy: The technical person writing the instructions should *not* be the cognizant engineer.**

The responsible design engineer is far too familiar with the situation to write the field change. A technical writer should draft the instructions and specify the kit, but a different tech should install one.

> **Policy: A technical person, other than the writer should incorporate the change into one product using the kit and instruction.**

How many of us have used a kit and instruction on Christmas eve to assemble a kids toy only to be frustrated by the instruction and end up with parts short or parts left over?

Having a person, different than the writer, install one kit will result in modifications and clarifications to the FCO which will help the field person when the factory is closed.

FIELD CHANGE FLOW

Thus the creation of the field change, assuming that the publications department is generating the FCO, should look like Figure 13.1.

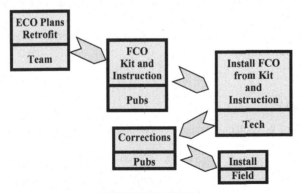

FIGURE 13.1 FCO flow.

The publications folks are probably the best technical writers available but the important factor is that the writer and the tech who does the trial installation are different people (preferably from different organizations) and neither should be the cognizant engineer.

FIELD CHANGE METRICS

The most basic of measurements for the field change process would be to measure the lapsed time to complete the FCO writing process, see Figure 13.2.

The time should be measured from the change order being technically released until the FCO is sent to the field. This analyst has no benchmark for how long this process should take. It would certainly vary with the complexity of the change but 3 weeks average is probably too long.

Every FCO should be tracked by field service to assure that each is fully installed. One way of tracking each one would be to graph and chart the progress, as shown in Figure 13.3.

Of course feedback from the field is required to track the retrofit. If you have online contact with the field service engineers it should be relatively easy to get feedback. If not, another possible method is to insert a self-addressed, stamped postcard in every kit.

CM should find out whether or not the field service folks are tracking the installation in some acceptable fashion.

FIELD CHANGE STANDARDS

Identify the policy and procedure for retrofit, repair, returns, and refurbishment, a subject this writer oversimplifies as "field change," changes after the initial shipment. The needed standards in this arena are as follows:

- **Field change policy: Which changes will be installed during retrofit, repair, etc.**
- **Field change form: Disassembly, reassembly, test, disposition of old parts, etc.**
- **Form instructions: Box-by-box form instruction or cursor pop-up instruction.**
- **Process flow diagram: Procedural sequence of events and responsibilities.**

SUMMARY

The important field change issues concerning CM and the senior management are as follows:

- Not all changes should be retrofit. Only noninterchangeable changes should be considered and not all of them should be retrofit.

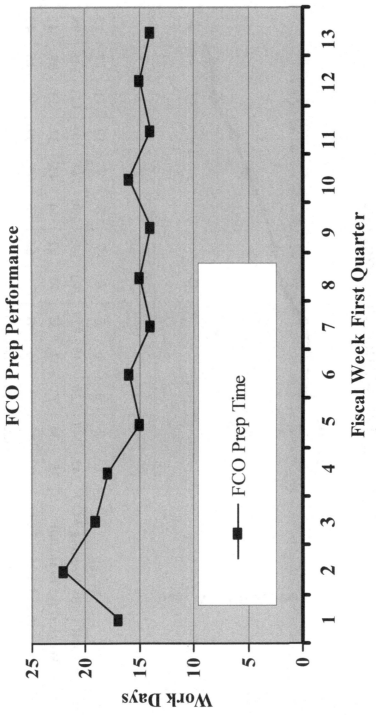

FIGURE 13.2 FCO preparation time.

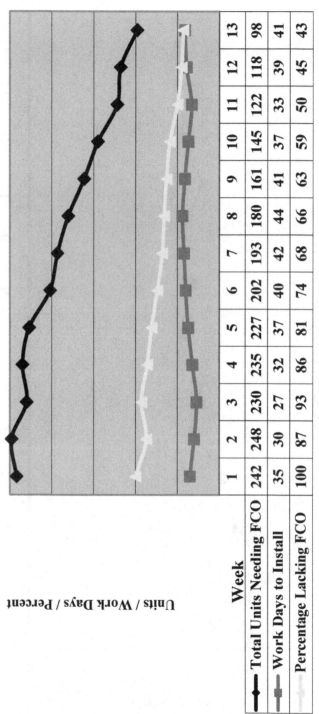

FIGURE 13.3 FCO installation tracking.

- The change process must include the FCO decisions in order to assure a wise choice as to which changes to retrofit and to determine the complete cost of the change.
- Recalls must be given extraordinary treatment.
- The kit of parts should contain a copy of the FCO and possibly a self-addressed, stamped postcard.
- The design engineer is not the best choice for FCO writing. A technician who is *not* familiar with the change should draft the field change and a different technician should install one to debug the kit and instruction.
- Always remember that when issues occur in the field, the factory is closed.

Chapter Fourteen

Take It to the Bank

You will recall that we identified the basic "raw materials" of product manufacturing—the very essence of the requirements for every product manufacturing operation—as follows:

Money—for start-up and from profits to prosper
Tools—building, machine, mold, software, etc.
People—and the policy/practices they choose
Product—embodied in design drawings, code, and specs

These four elements must be present and robust in any successful product manufacturing company.

We hopefully have established that the management of the design documents is a critical company discipline. Without precise and controlled design documents, we do not have a producible product. Without minimum control of design documents with make-sense processes, practices, and measurements, you will have some degree of chaos and waste.

Besides the imperative nature of the discipline, there are huge benefits for doing "best-in-class" configuration management.

BENEFITS

CM, kept simple, results in many benefits to the company. The benefits of make sense, documented, fast, accurate, efficient, effective, measured, well-understood, minimally controlled, process approach to CM system are plentiful.

- Gets new products to the market faster.
- Reduces promise to deliver time.
- Happier customers because they see the new option, change, or feature they had requested, much quicker.
- The customers get what they ordered in the delivery time promised.
- Reduces the "bone piles" of down-level material that will probably eventually be scrapped.
- Gets real cost reductions implemented quicker.

157

Configuration Management for Senior Managers. http://dx.doi.org/10.1016/B978-0-12-802382-2.00014-3

- Reduces the manufacturing rework and scrap costs significantly.
- Improves bill of material (BOM) accuracy and saves the corresponding material waste and correction time.
- Improvement in part, assembly, specification, and product quality.
- Eliminates multiple BOMs and saves the costs of maintaining the bills, not to mention eliminating the risks associated with multiple bills.
- Evolution of the BOM in lead time to produce the product quicker.
- Reduces field maintenance, retrofit, and repair cost.
- Know exactly what items are noninterchangeable in each product.
- Improves the understanding and communication between design engineering and the rest of the world.
- Clarifies responsibilities and thus eliminates finger pointing.
- Saves wear and tear on CM managers, master schedulers, and all types of engineers.
- Complies with applicable customer and agency standards.
- Sorts out changes that are not needed or aren't cost-effective.
- Saves many dollars a year in paper and copying costs alone.
- Significant reduction in the cost of quality.
- Allows the company to qualify as a best-in-class producer.
- Sets the stage for innovation in engineering and operations.

The ways and means of achieving these benefits are not secret, high-tech, or cost-prohibitive. These benefits are attainable by following the outline in this book.

CRITICAL TO SUCCESS

A European graduate student asked the writer to delineate the critical success factors for a CM process implementation and improvement. The following is the result (edited) of this author's response.

It presumes that we were starting with a functioning company with some kind of processes in place. The most critical CM factors to success—after defining which process needs improvement and without trying to improve them all at once—are as follows:

If reinventing or reengineering a process:

- **Executive management champion**—The chief engineer, a VP, or general manager must be dedicated to the success of the project. This requires recognition of the problem(s) and a fervent desire to see a complete redesign of the process.
- **Recognition of the problems** and recognition of the fact that many months of time on the part of a few people will be required.
- **All other elements listed under continuous improvement** are also critical—especially the work flow diagram.
- Approval to obtain **outside consultation** if needed.

Lacking such a management champion or the dedication of time from a few team members, continuous improvement is the best path of success.

If doing continuous improvement of a process:

- Some management backing.
- A "missionary" CM manager is required. That person must be in charge of the CM function and have a dogged tenacity about improving a CM process. Few CM managers are not aware of many of the problems needing improvement. Only a few have the fervent desire and the time to do the job.
- A CM manager with at least half time to spend on the improvement project.
- Key measurements and work flow diagram of the current process are put in place first.
- Ideally (but not mandatory) half time from three other people to spend on the project. One from operations (preferably production control), one from engineering, and one from the supply chain under the CM managers direction.
- Approval to obtain outside consultation help if needed.
- Write the standards for the general foundation blocks, circulate, sell, revise, recirculate, sell, sign, and implement.
- After all the general standards are in place, the process flow can be addressed and changes made in small bites.
- Definition of the problem(s) for one process. A list of the goals for that improvement project. Prioritize the list—easiest to accomplish on the top.
- Design an improved work flow goal.
- The improvements are done in very small bites, starting with standards for that process and working toward a streamlined work flow.

Always be guided by the

Platinum policy—Any process invented by man can be improved by man.

Index

Note: Page numbers followed by "f" and "t" indicate figures and tables, respectively.

Printed in the United States
By Bookmasters